Math Mammoth
Grade 1-B Worktext

By Maria Miller

Contents

Chapter 7: Adding and Subtracting Within 0-100

Chapter 8: Coins

Foreword

Math Mammoth Grade 1 comprises a complete math curriculum for the first grade mathematics studies. The curriculum meets and exceeds the Common Core standards.

The main areas of study for first grade are:

1. The concepts of addition and subtraction, and strategies for addition and subtraction facts;

2. Developing understanding of place value up to 100;

3. Developing understanding and some basic strategies for two-digit addition and subtraction.

Additional topics we study in the first grade are telling time (whole and half hours), geometric shapes, measurement, and counting coins.

The portion of first grade included in this book, Part B, covers strategies for addition and subtraction facts (chapter 4), telling time and the calendar (chapter 5), shapes and measuring (chapter 6), adding and subtracting two-digit numbers and reading pictographs (chapter 7), and counting coins (chapter 8). The book 1-A covers addition concept, subtraction concept, and place value with two-digit numbers.

Some important points to keep in mind when using the curriculum:

• These two books (parts A and B) are like a "framework", but you still have a lot of liberty in planning your child's studies. While addition and subtraction topics are best studied in the order they are presented, feel free to go through the sections on shapes, measurement, clock, and money in any order you like.

   This is especially advisable if your child is either "stuck" or is perhaps getting bored with some particular topic. Sometimes the concept the child was stuck on can become clear after a break from the topic.

• Math Mammoth is mastery-based, which means it concentrates on a few major topics at a time, in order to study them in depth. However, you can still use it in a *spiral* manner, if you prefer. Simply have your child study in 2-3 chapters simultaneously. This type of flexible use of the curriculum enables you to truly individualize the instruction for your child.

• Don't automatically assign all the exercises. Use your judgment, trying to assign just enough for your child's needs. You can use the skipped exercises later for review. For most children, I recommend to start out by assigning about half of the available exercises. Adjust as necessary.

• For review, the curriculum includes a worksheet maker (Internet access required), mixed review lessons, additional cumulative review lessons, and the word problems continually require usage of past concepts. Please see more information about review (and other topics) in the FAQ at **https://www.mathmammoth.com/faq-lightblue.php**

I heartily recommend that you view the full user guide for your grade level, available at **https://www.mathmammoth.com/userguides/**

There are free videos matched to the curriculum at **https://www.mathmammoth.com/videos/**

I wish you success in teaching math!

Maria Miller, the author

Chapter 4: Addition and Subtraction Facts
Introduction

This chapter provides lots of practice for learning and memorizing the basic addition and subtraction facts with numbers from 0 to 10. The Common Core Standards require children in the first grade to demonstrate fluency in addition and subtraction with numbers up to 10, and we aim for that goal here.

Since this chapter is repetitive, consider studying it simultaneously with some other section of the curriculum, such as telling time, shapes, measuring, or counting coins. For example, the child could study telling time and this chapter each day, or study the two different chapters on alternate days. This is not compulsory but just a suggestion to "mix things up" in a somewhat spiral fashion.

The series of lessons titled *Addition and Subtraction Facts with...* aims to help the student to memorize the basic addition and subtraction facts within 0-10. We approach it from the concept of "fact families," which makes the process logical and structured. These lessons have a lot of repetition and practice for both subtraction and addition facts.

Many children may not need all the practice problems provided, so don't assign all of them by default. Use your judgment, and only assign a certain portion, such as half of them, at first. The rest of them can then be used later as a review. If assigning only half of the exercises is not enough, adjust as necessary.

Alongside this book, you can also use math games or flashcards to reinforce these facts. You will find a list of some games below.

While the child does not absolutely have to learn these facts by heart while studying this chapter, it is advisable to learn them fairly well. Mathematics builds upon previously learned concepts and facts, and learning addition and subtraction facts is essential for later study, such as when students add $24 + 5$ (in chapter 7 of this curriculum). However, if the child has not memorized these facts before the end of the chapter, don't worry. Go on with the curriculum, but keep practicing the facts with games, worksheets, drills, *etc.*

Besides practicing the facts with the help of fact families, the child will also solve word problems, fill in number patterns, use a symbol to represent an unknown number, compare expressions (such as $5 - 2 < 2 + 5$), and subtract more than one number at a time.

As a friendly reminder, there are videos matched to the curriculum at https://www.mathmammoth.com/videos. Choose Grade 1.

Pacing Suggestion for Chapter 4

The Lessons in Chapter 4	page	span	suggested pacing	your pacing
Addition and Subtraction Facts with 4 and 5	10	*2 pages*	2 days	
Addition and Subtraction Facts with 6	12	*3 pages*	2 days	
Addition and Subtraction Facts with 7	15	*2 pages*	2 days	
Addition and Subtraction Facts with 8	17	*4 pages*	3 days	
Addition and Subtraction Facts with 9	21	*3 pages*	2 days	
Addition and Subtraction Facts with 10	24	*4 pages*	3 days	
Subtracting More Than One Number	28	*2 pages*	1 day	
Review - Facts with 6, 7, and 8	30	*2 pages*	1 day	
Review - Facts with 9 and 10	32	*3 pages*	2 days	
Chapter 4 Test (optional)				
TOTALS		*25 pages*	18 days	

Please add one day to the pacing for the test if you will use it. Note that the specific lessons in the chapter can take several days to finish. They are not "daily lessons." As a general guideline, first graders should finish 1-2 pages daily or 7-9 pages a week. Please also see the user guide at https://www.mathmammoth.com/userguides/ .

Games and Activities

Addition (or Subtraction) Challenge

You need: A standard deck of playing cards from which you remove the face cards. For the addition version, you might also remove some of the other higher cards, such as tens, nines, and eights.

Game Play: In each round, each player is dealt two cards face up, and has to calculate their sum or difference (add/subtract). The player with the highest sum or difference gets all the cards from the other players. After enough rounds have been played to use all of the cards, the player with the most cards wins. If two or more players have the same sum, then those players get an additional two cards and use those to resolve the tie.

Number Bonds in the Pond

You need: A standard deck (or several) of playing cards or number cards

Preparation: Choose a target sum for the game. If the target sum is 5, make a deck of cards consisting of numbers 1 through 4. If the target sum is 6, make a deck of numbers 1-5. And so on. (The deck always consists of numbers that are from 1 through $X - 1$ where X is the target sum.) Place a target number card face up between the players, and spread out the rest of the cards face down, like a pond, between the players.

Game play: At your turn, if you don't have any cards in your hand, take two cards from the pond. If you do, take one card from the pond. Now check if any two cards in your hand add up to the target number. If so, put those cards away to your personal pile. If not, it is the next player's turn. The game ends when there are no more cards in the pond. The winner is the person with most cards in their personal pile.

Variation: Allow three cards/numbers to be added to reach the target number.

Notes: Depending on the number of players, you may need several decks of cards to make the pond. Playing this game several times will help the child to memorize the number bonds associated with a particular target number.

10 Out (or *6 Out, 7 Out, 8 Out, etc.*)

You need: A deck of number cards with numbers 1-10, or regular playing cards without the face cards.

Preparation: Choose a target sum, such as 10. Deal seven cards to each player. Place the rest face down in a pile in the middle of the table.

Game play: At your turn, first take one card from the pile. Then try to find pairs of cards in your hand that add up to 10, and discard any such pairs. Discard the card 10 also if you have it. If you cannot find any such pairs, ask for any one card you want (such as 6) from the player to your right (as in "Go Fish"). That player, if he has it, must give it, and you will then discard the pair that makes 10. Then it is the next player's turn. The player who first discards all the cards from his hand is the winner.

Variations:
* Deal more than seven cards.
* Deal fewer cards if there are a lot of players or the players are very young.
* Allow players to discard *three* cards that add up to 10.
* Instead of ten, players discard cards that add up to 5, 6, 7, 8, or 9.

Games and Activities at Math Mammoth Practice Zone

Fact Families
Choose which fact family or families to practice, and the program will give you addition and subtraction problems from those, including with missing numbers.
https://www.mathmammoth.com/practice/fact-families

Subtraction Hidden Picture Game
Choose a number range (such as 1 to 10) and uncover a hidden picture while solving subtraction problems!
https://www.mathmammoth.com/practice/mystery-picture-subtraction

Number Bonds
Practice number bonds, either with pictures or with numbers.
https://www.mathmammoth.com/practice/number-bonds

"7 Up" Card Game
You will see seven cards dealt face up. Simply choose any two cards that make 10 (or your chosen sum) to discard. When there are no cards that make that sum, click the deck to deal more cards. For this chapter, choose sums of 7, 8, 9, or 10.
https://www.mathmammoth.com/practice/seven-up

Fruity Math
Add two single-digit numbers (such as 4 + 5). Click the fruit with the correct answer and try to get as many points as you can within two minutes.
https://www.mathmammoth.com/practice/fruity-math#op=addition&duration=120&mode=manual&config=2,5x1__1,7x1&max-sum=120

Bingo
Choose Subtraction (Under 10).
https://www.mathmammoth.com/practice/bingo

Make Subtraction Sentences
You are given numbers (in flowers), and an answer to a subtraction. Drag two flowers to the empty slots so that the subtraction is true.
https://www.mathmammoth.com/practice/number-sentences#questions=5&types=sub-1-12

Further Resources on the Internet

We have compiled a list of Internet resources that match the topics in this chapter, including pages that offer:

- **online practice** for concepts;
- online **games**, or occasionally, printable games;
- **animations** and interactive **illustrations** of math concepts;
- **articles** that teach a math concept.

We heartily recommend you take a look! Many of our customers love using these resources to supplement the bookwork. You can use these resources as you see fit for extra practice, to illustrate a concept better and even just for some fun. Enjoy!

https://l.mathmammoth.com/gr1ch4

Scan me

Addition and Subtraction Facts with 4 and 5

Facts with 4		$4 + 0 = 4$ $\quad$ $0 + 4 = 4$	$4 - 4 = 0$ $\quad$ $4 - 0 = 4$
		$1 + 3 = 4$ $\quad$ $3 + 1 = 4$	$4 - 3 = 1$ $\quad$ $4 - 1 = 3$
		$2 + 2 = 4$	$4 - 2 = 2$

Facts with 5		$5 + 0 = 5$ $\quad$ ___ + ___ = 5	$5 - 5 = 0$ $\quad$ $5 -$ ___ = ___
		$4 + 1 = 5$ $\quad$ $1 + 4 = 5$	$5 - 4 =$ ___ $\quad$ $5 -$ ___ = ___
		$3 + 2 = 5$ $\quad$ ___ + ___ = 5	$5 - 3 =$ ___ $\quad$ $5 -$ ___ = ___

1. Find the missing numbers.

a.	b.	c.	d.
$3 +$ ___ $= 4$	$2 +$ ___ $= 5$	$5 - 0 =$ ___	$4 - 0 =$ ___
$1 +$ ___ $= 4$	$1 +$ ___ $= 5$	$5 - 4 =$ ___	$4 - 3 =$ ___
$1 +$ ___ $= 5$	$4 +$ ___ $= 5$	$5 - 2 =$ ___	$5 - 1 =$ ___
$2 +$ ___ $= 4$	$3 +$ ___ $= 5$	$4 - 1 =$ ___	$4 - 2 =$ ___

2. Color the square:

  - yellow if the answer is 0,
  - red if the answer is 1,
  - blue if the answer is 2,
  - green if the answer is 3,
  - purple if the answer is 4,
  - orange if the answer is 5.

$5-4$	$2+3$	$4-4$	$1+2$	$4-2$	$1+3$
$2+2$	$3-2$	$5-0$	$0+0$	$5-2$	$1+1$
$0+2$	$5-1$	$0+1$	$1+4$	$0-0$	$4-1$

3. Continue the patterns until the boxes are full!

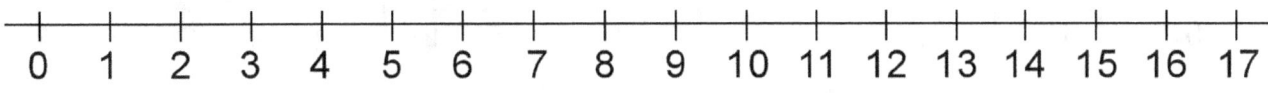

	a.	b.	c.

a.

$17 - 0 = $ _____

$17 - 1 = $ _____

$17 - 2 = $ _____

$17 - $ ____ $ = $ _____

$17 - $ ____ $ = $ _____

____ $ - $ ____ $ = $ _____

____ $ - $ ____ $ = $ _____

b.

$10 + $ ____ $ = 10$

$10 + $ ____ $ = 11$

$10 + $ ____ $ = 12$

$10 + $ ____ $ = $ ____

____ $ + $ ____ $ = $ ____

____ $ + $ ____ $ = $ ____

____ $ + $ ____ $ = $ ____

c.

$5 - 2 = $ _____

$6 - 2 = $ _____

$7 - 2 = $ _____

____ $ - 2 = $ _____

____ $ - $ ____ $ = $ _____

____ $ - $ ____ $ = $ _____

____ $ - $ ____ $ = $ _____

Addition and Subtraction Facts with 6

1. Complete the fact families in which the sum is six. At the top, write the three numbers that you are using for the fact family.

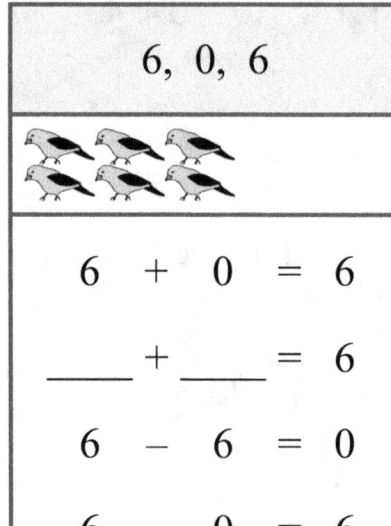

6, 0, 6

6 + 0 = 6

____ + ____ = 6

6 – 6 = 0

6 – 0 = 6

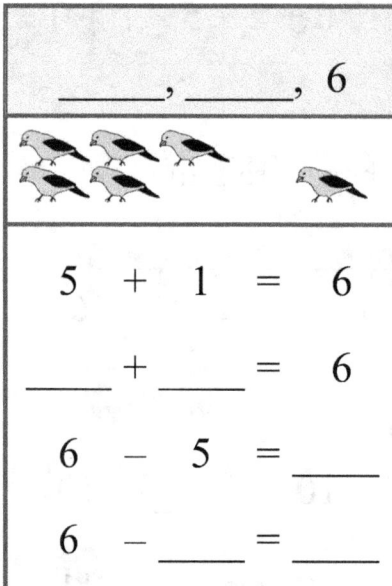

____, ____, 6

5 + 1 = 6

____ + ____ = 6

6 – 5 = ____

6 – ____ = ____

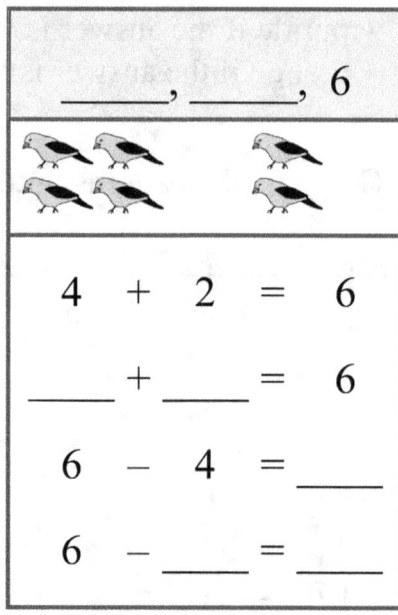

____, ____, 6

4 + 2 = 6

____ + ____ = 6

6 – 4 = ____

6 – ____ = ____

____, ____, 6

3 + 3 = 6

6 – 3 = ____

2. Write the numbers that add up to 6. Memorize these!

0 + ____ = 6   or   ____ + 0 = 6

1 + ____ = 6   or   ____ + 1 = 6

2 + ____ = 6   or   ____ + 2 = 6

3 + ____ = 6

3. Subtract.

a.   6   b.   6   c.   6   d.   6   e.   6   f.   6

   – 5   – 4   – 6   – 2   – 1   – 3

4. Play the "6 Out" card game (see p. 7, in the introduction to this chapter).

5. Find the missing numbers.

a.	b.	c.	d.
4 + ____ = 6	5 + ____ = 6	____ + 2 = 6	6 – ____ = 2
3 + ____ = 6	0 + ____ = 6	____ + 1 = 6	6 – ____ = 5

6. For each "how many more" addition, write a subtraction using the same numbers so that the numbers in the triangles are the same.

a. 2 + △ = 5

__5__ – 2 = △3

b. 1 + △ = 6

____ – ____ = △

c. 4 + △ = 5

____ – ____ = △

d. 3 + △ = 8

__8__ – 3 = △

e. 5 + △ = 10

____ – ____ = △

f. 2 + △ = 7

____ – ____ = △

7. Write fact families so that the numbers in the triangles are the same.

a. 1 + △ = 6

△ + ____ = 6

____ – △ = ____

____ – ____ = △

b. 2 + 7 = △

____ + ____ = △

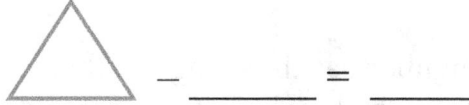

△ – ____ = ____

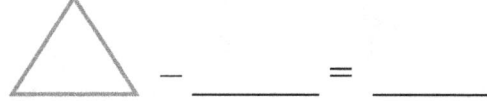

△ – ____ = ____

13

8. Solve the word problems. Think:

- Are you asked for the total? $(2 + 4 = ?)$ OR

- Are you asked how many more? $(2 + ? = 4)$ OR

- Are you asked how many are left? $(4 - 2 = ?)$

---

a. The black cat has four kittens and the white cat has three.
How many kittens do they have altogether?

How many more kittens does the black cat have than the white cat?

---

b. John had ten crayons but now he only has two.
How many crayons has he lost?

---

c. Mother found ten clothespins in one container and two in another.
How many clothespins were in the two containers?

How many more clothespins were in the first container
than in the second?

---

d. Jill has two eggs. She needs eight eggs to make some
cakes. How many more eggs does Jill need?

Her neighbor has three eggs. If the neighbor gives Jill the
three eggs she has, how many more does Jill still need?

Addition and Subtraction Facts with 7

1. Complete the fact families with 7. At the top, write the three numbers you are using.

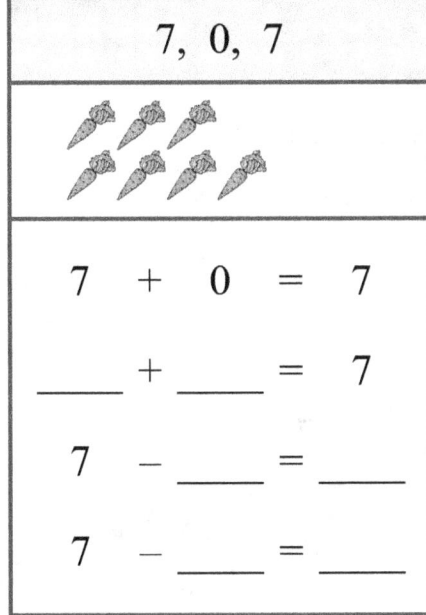

7, 0, 7

$$7 + 0 = 7$$

___ + ___ = 7

$$7 - \underline{\quad} = \underline{\quad}$$

$$7 - \underline{\quad} = \underline{\quad}$$

___, ___, 7

$$6 + \underline{\quad} = 7$$

___ + ___ = 7

$$7 - \underline{\quad} = \underline{\quad}$$

___ − ___ = ___

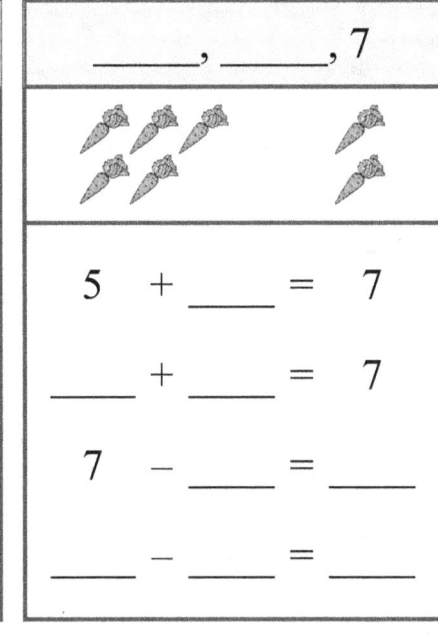

___, ___, 7

$$5 + \underline{\quad} = 7$$

___ + ___ = 7

$$7 - \underline{\quad} = \underline{\quad}$$

___ − ___ = ___

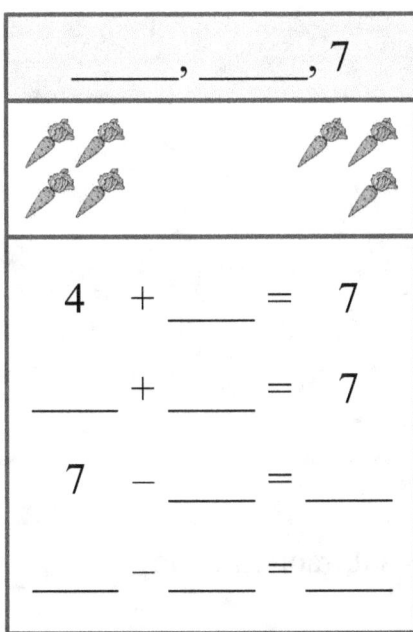

___, ___, 7

$$4 + \underline{\quad} = 7$$

___ + ___ = 7

$$7 - \underline{\quad} = \underline{\quad}$$

___ − ___ = ___

2. Write down the numbers that add up to 7 and memorize the addition facts!

0 + ___ = 7	or	___ + 0 = 7
1 + ___ = 7	or	___ + 1 = 7
2 + ___ = 7	or	___ + 2 = 7
3 + ___ = 7	or	___ + 3 = 7

3. Play the "7 Out" card game (see p. 7, in the introduction to this chapter).

4. Subtract.

a. 7
   − 5
   ___

b. 7
   − 4
   ___

c. 7
   − 6
   ___

d. 7
   − 2
   ___

e. 7
   − 1
   ___

f. 7
   − 3
   ___

5. Fill in. Then draw a line between the facts that are from the same fact family.

5 + ____ = 7	7 − ____ = 4	7 − ____ = 6
7 − 0 = ____	6 + ____ = 7	____ + 4 = 7
7 − 3 = ____	7 − 2 = ____	7 − ____ = 7
7 − ____ = 1	0 + ____ = 7	7 − ____ = 2

6. Solve.

a. Luis has 4 pencils and Jeremy has 6.
How many more pencils does Jeremy have than Luis?

How many pencils do the two boys have altogether?

b. Maria found two socks in the hamper, five socks in her basket, and one sock on the floor. How many socks did she find?

Puzzle Corner Figure out how to fill in the rest of this subtraction table?

−	12	11	10	9	8	7	6	5	4	3
1	11					6				
2		9					4			1

Addition and Subtraction Facts with 8

1. Complete the fact families in which the sum is eight.

8, _____, 8

8 + 0 = 8

_____ + _____ = 8

8 − _____ = _____

_____ − _____ = _____

_____, _____, 8

7 + 1 = 8

_____ + _____ = 8

8 − _____ = _____

_____ − _____ = _____

_____, _____, 8

_____ + _____ = _____

_____ + _____ = _____

_____ − _____ = _____

_____ − _____ = _____

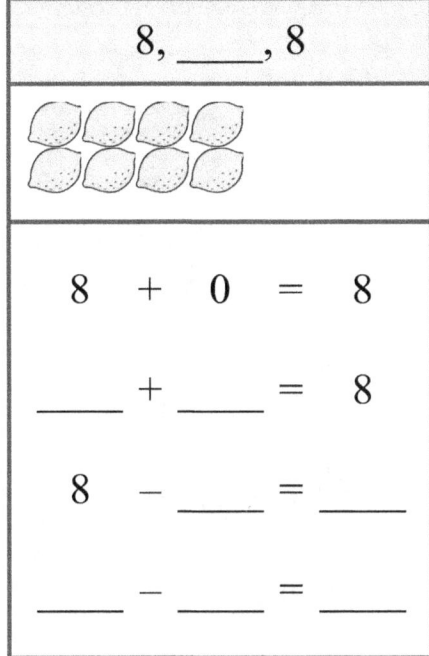

_____, _____, 8

_____ + _____ = 8

_____ + _____ = 8

_____ − _____ = _____

_____ − _____ = _____

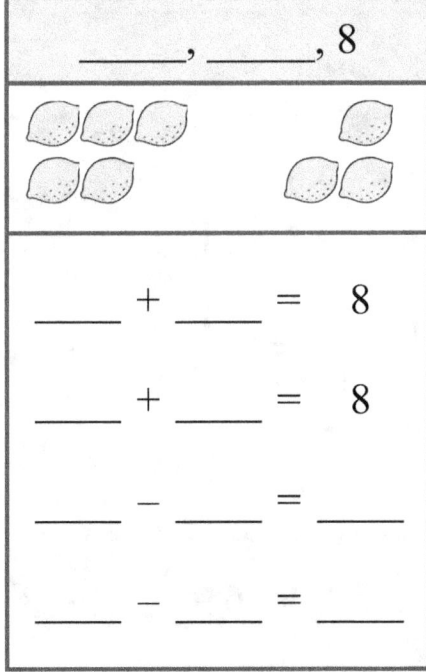

_____, _____, 8

_____ + _____ = _____

_____ − _____ = _____

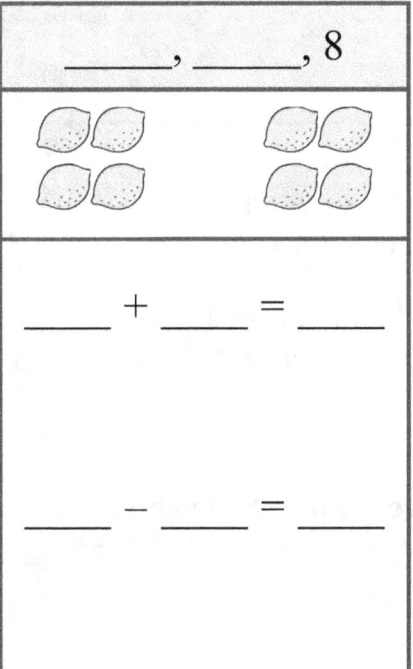

2. Play the "8 Out" card game (see p. 7, in the introduction to this chapter).

3. Write the addition facts with 8 and memorize them!

0 + _____ = 8   or   _____ + 0 = 8   4 + _____ = 8

1 + _____ = 8   or   _____ + 1 = 8

2 + _____ = 8   or   _____ + _____ = 8

3 + _____ = 8   or   _____ + _____ = 8

4. Find the missing numbers.

a.	b.	c.	d.
8 − 3 = _____	5 + _____ = 8	4 + _____ = 8	8 − 7 = _____
8 − 1 = _____	1 + _____ = 8	6 + _____ = 8	8 − 0 = _____
8 − 2 = _____	2 + _____ = 8	7 + _____ = 8	8 − 4 = _____

First subtract 6 – 2 to get 4.
Write 4 in the shaded box below.

Then compare 4 and 5. Since
four is less than five, write "<".

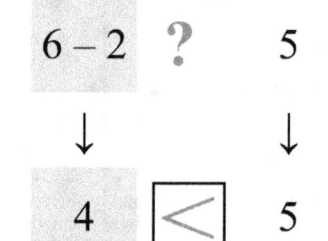

5. First subtract. Write the answer in the shaded box below.

a. 5 − 2 ? 4	b. 7 − 4 ? 5	c. 8 − 1 ? 7	d. 6 − 3 ? 2
↓ ↓	↓ ↓	↓ ↓	↓ ↓
☐ 4	☐ 5	☐	☐

First find $4 + 2$ (on the left side) and $8 - 3$ (on the right side).

Write the answers below. Then compare. Which is greater? Six is greater. So write ">" in the box.

$$4 + 2 \quad ? \quad 8 - 3$$
$$\downarrow \qquad\qquad \downarrow$$
$$6 \quad \boxed{>} \quad 5$$

6. First add and subtract. Write the answer in the box below. Then compare, and write $<$, $>$ or $=$.

a. $5 - 2 \quad ? \quad 4 - 2$	**b.** $8 - 1 \quad ? \quad 7 - 1$	**c.** $8 - 6 \quad ? \quad 8 - 5$
d. $6 + 2 \quad ? \quad 7 + 2$	**e.** $7 - 1 \quad ? \quad 7 - 2$	**f.** $4 + 4 \quad ? \quad 7 - 5$
g. $1 - 1 \quad ? \quad 3 - 2$	**h.** $3 + 10 \quad ? \quad 10$	**i.** $7 \quad ? \quad 4 + 2$
j. $8 - 1 \quad ? \quad 4 - 2$	**k.** $7 - 2 \quad ? \quad 6 - 1$	**l.** $9 - 0 \quad ? \quad 7 + 2$

7. Fill in the missing numbers. Then draw a line between the facts that are from the same fact family.

6 + _____ = 8

8 − 0 = _____

8 − 3 = _____

_____ + 1 = 8

4 + _____ = 8

8 − 4 = _____

8 − _____ = 3

7 + _____ = 8

8 − 2 = _____

0 + _____ = 8

_____ + 5 = 8

8 − _____ = 6

_____ + 4 = 8

8 − _____ = 8

8 − 1 = _____

8. Solve the word problems. Drawing can help you.

a. Jack has 10 cars, Bill has 7, and Ed has 4.

How many more cars than Ed does Bill have?

How many more cars than Ed does Jack have?

How many more cars than Bill does Jack have?

b. Mary has saved seven dollars. She would like to buy a puzzle for five dollars and a game for three dollars.
Can she buy both things?

If she can, how much money does she have left over?

If she can't, how much more money would she need to save?

Addition and Subtraction Facts with 9

1. Write the fact families where the sum is 9.

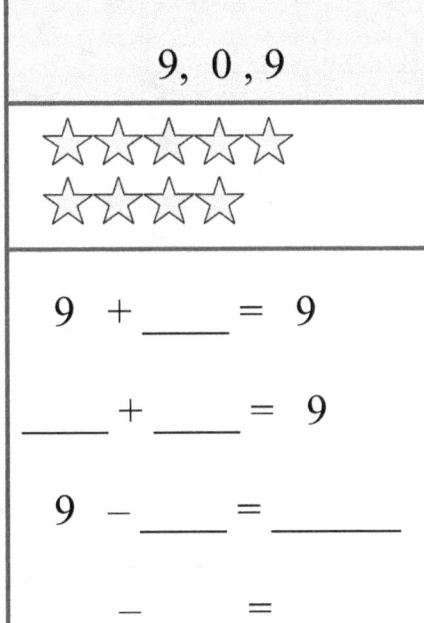

9, 0 , 9

9 + ____ = 9

____ + ____ = 9

9 − ____ = ____

____ − ____ = ____

____ , ____ , 9

____ + ____ = ____

____ + ____ = ____

____ − ____ = ____

____ − ____ = ____

____ , ____ , 9

____ + ____ = ____

____ + ____ = ____

____ − ____ = ____

____ − ____ = ____

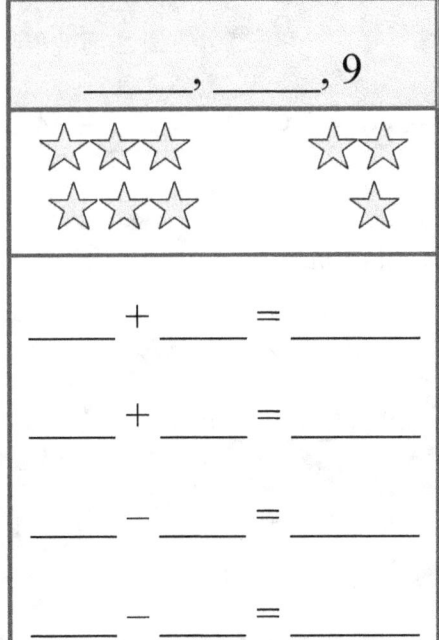

____ , ____ , 9

____ + ____ = ____

____ + ____ = ____

____ − ____ = ____

____ − ____ = ____

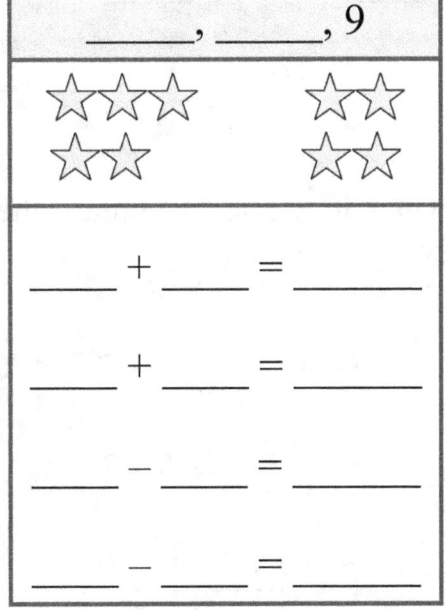

____ , ____ , 9

____ + ____ = ____

____ + ____ = ____

____ − ____ = ____

____ − ____ = ____

2. Play the "9 Out" card game (see p. 7, in the introduction to this chapter).

3. Write the addition
   facts with 9 and
   memorize them!

0 + _____ = 9   or   _____ + 0 = 9

1 + _____ = 9   or   _____ + 1 = 9

2 + _____ = 9   or   _____ + _____ = 9

3 + _____ = 9   or   _____ + _____ = 9

4 + _____ = 9   or   _____ + _____ = 9

4. Find the missing numbers.

a.	b.	c.	d.
9 – 5 = _____	2 + _____ = 9	_____ + 1 = 9	9 – _____ = 1
9 – 3 = _____	1 + _____ = 9	_____ + 3 = 9	9 – _____ = 0
9 – 6 = _____	7 + _____ = 9	_____ + 5 = 9	9 – _____ = 2
9 – 8 = _____	8 + _____ = 9	_____ + 7 = 9	9 – _____ = 4

5. Fill in the missing numbers. Then draw lines to connect the facts that
   belong to the same fact family.

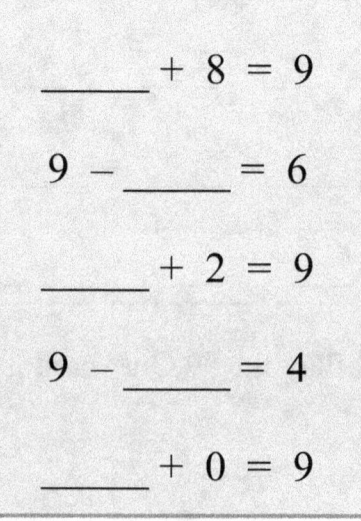

7 + _____ = 9

9 – 3 = _____

9 – 1 = _____

9 – _____ = 9

4 + _____ = 9

0 + _____ = 9

9 – _____ = 3

5 + _____ = 9

9 – 2 = _____

9 – _____ = 1

_____ + 8 = 9

9 – _____ = 6

_____ + 2 = 9

9 – _____ = 4

_____ + 0 = 9

22

6. First add or subtract. Don't write answers in the box—just solve them in your head!
   Then compare and write $<$, $>$ or $=$ in the box.

a.   8 ☐ $10-3$ b.   9 ☐ $9+3$ c. $8-6$ ☐ $6+3$

d. $6+2$ ☐ $8+2$ e. $10-1$ ☐ 10 f. $8-4$ ☐ $8-5$

g. $5-2$ ☐ $4-2$ h. $8+0$ ☐ $8-0$ i. $9-1$ ☐ $9+1$

7. Subtract.

a.	b.	c.	d.	e.	f.
9	9	9	8	9	8
-5	-4	-6	-2	-2	-3

8. If the answer is 6 or 7, color its box blue. If the answer is 8 or 9, color its box red,
   and color the rest of the boxes yellow.

$9-3$	$4+6$	$9-0$	$4+6$	$8-1$
$2+5$	$9-5$	$4+4$	$4-2$	$5+1$
$9-2$	$3+7$	$10-2$	$10+0$	$7-1$
$4+2$	$7-3$	$6+3$	$3-1$	$3+3$
$6-0$	$1+1$	$8-0$	$3+2$	$10-4$
$3+4$	$8-3$	$2+7$	$7-6$	$7+0$
$1+6$	$2+8$	$10-1$	$2+2$	$7-0$

Addition and Subtraction Facts with 10

1. Complete the fact families in which the sum is ten.

10, 0, 10	9, 1, 10	_____, _____, 10
10 + 0 = 10	9 + ____ = 10	8 + ____ = 10
____ + ____ = 10	____ + ____ = 10	____ + ____ = 10
10 – 0 = ____	10 – 9 = ____	10 – ____ = ____
10 – ____ = ____	10 – ____ = ____	10 – ____ = ____

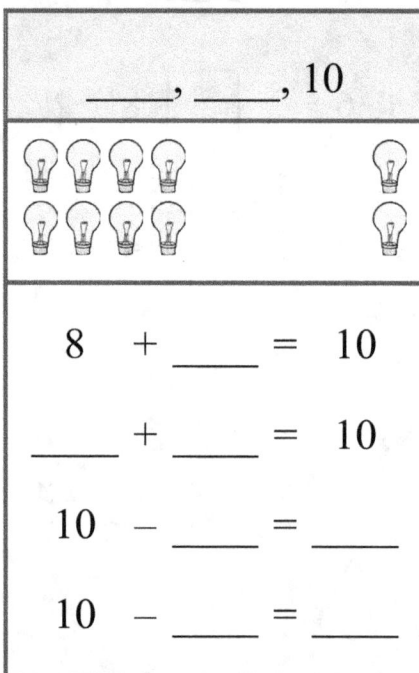

_____, _____, 10	_____, _____, 10	_____, _____, 10
7 + ____ = 10	____ + ____ = ____	____ + ____ = ____
____ + ____ = 10	____ + ____ = ____	____ + ____ = ____
10 – ____ = ____	____ – ____ = ____	____ – ____ = ____
10 – ____ = ____	____ – ____ = ____	____ – ____ = ____

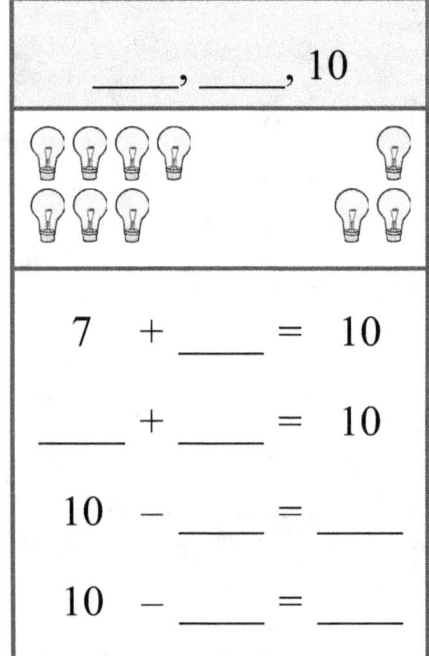
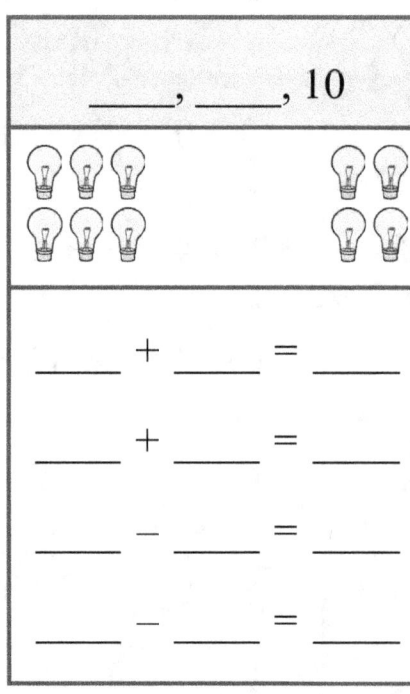
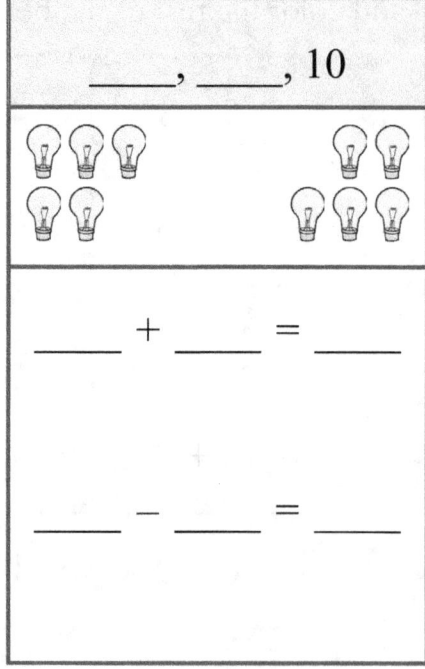

2. Play the "10 Out" card game (see p. 7, in the introduction to this chapter).

3. Write the
   addition facts
   with 10 and
   memorize them!

$0 +$ _____ $= 10$ or   _____ $+ \; 0 \; = 10$

$1 +$ _____ $= 10$ or   _____ $+ \; 1 \; = 10$

$2 +$ _____ $= 10$ or   _____ $+$ _____ $= 10$

$3 +$ _____ $= 10$ or   _____ $+$ _____ $= 10$

$4 +$ _____ $= 10$ or   _____ $+$ _____ $= 10$

$5 +$ _____ $= 10$

4. Find the missing numbers.

a.	b.	c.
$10 - 3 =$ _____	$0 +$ _____ $= 10$	$10 -$ _____ $= 7$
$10 - 6 =$ _____	$7 +$ _____ $= 10$	$10 -$ _____ $= 2$
$10 - 2 =$ _____	$9 +$ _____ $= 10$	$10 -$ _____ $= 4$

5. Fill in. Then draw lines to connect the facts that are from the same fact family.

$6 +$ _____ $= 10$	$0 +$ _____ $= 10$	_____ $+ 5 = 10$
$10 - 5 =$ _____	$8 +$ _____ $= 10$	$10 -$ _____ $= 6$
$10 - 1 =$ _____	$10 -$ _____ $= 3$	_____ $+ 2 = 10$
$10 - 2 =$ _____	$5 +$ _____ $= 10$	$10 -$ _____ $= 7$
$10 -$ _____ $= 10$	$10 - 4 =$ _____	_____ $+ 10 = 10$
$7 +$ _____ $= 10$	$9 +$ _____ $= 10$	$10 - 9 =$ _____

6. Write an addition sentence and a subtraction sentence for these word problems.
   Remember, you can always draw a picture of the situation to help you!

a. Sarah has six coins in her piggy bank. Elisa has two coins in hers.

Today Elisa found three coins on the ground.

Now who has more coins?

How many more?

b. Dad had one box of nails at home, and then he bought six more boxes of nails.
The next day he gave one box to the neighbor.

How many boxes of nails does Dad have now?

c. You have six dollars and I have three.
Together can we buy a meal that costs $8?

If so, is there any money left over, and how much?

If not, how much more money would we need?

7. More missing numbers!

a. $10 - \boxed{} = 5$ **b.** $3 + \boxed{} = 10$ **c.** $10 - \boxed{} = 6$

$10 - \boxed{} = 7$ $4 + \boxed{} = 10$ $10 - \boxed{} = 7$

$10 - \boxed{} = 2$ $1 + \boxed{} = 10$ $10 - \boxed{} = 9$

8. Several girls counted how many coins they had. Make a bar graph. This means you need to draw the bars. Just draw rectangles so that they reach the given numbers.

Girl	Coins
Lisa	9
Maggie	11
Lily	6
Susana	8

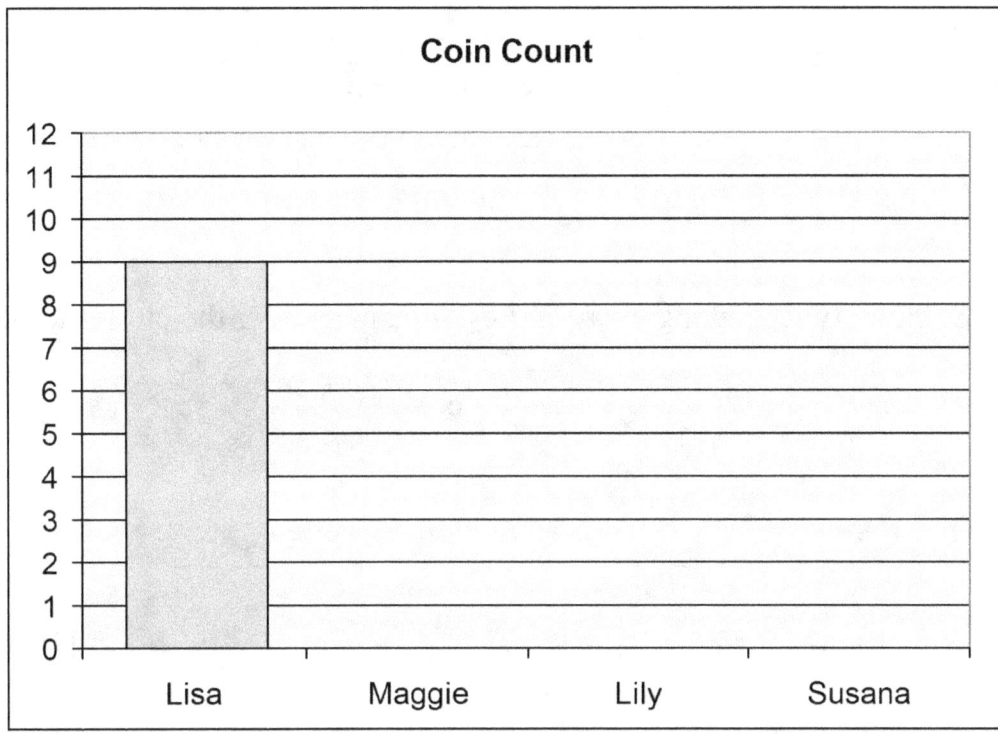

9. Make up two "how many more" questions about the bar graph. Ask a friend the questions. Check the student's answers!

Puzzle Corner Let △ represent one number and ☐ represent another number. They are different in each case (**a**, **b**, and **c**). For each case, find what the numbers are. Just guess and check!

a.

☐ + △ = 10

☐ − △ = 2

b.

☐ + △ = 10

☐ − △ = 6

c.

☐ + △ = 10

☐ − △ = 0

Subtracting More Than One Number

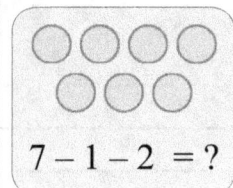

$7 - 1 - 2 = ?$

You have 7 circles. First you take away 1 circle, and then you take away 2 more circles.

You will have 4 circles left.

1. Subtract twice, taking away circles. You can cover the circles to help.

a.	b.	c.
$8 - 2 - 3 =$ ____	$9 - 3 - 1 =$ ____	$10 - 5 - 3 =$ ____
$8 - 5 - 2 =$ ____	$9 - 4 - 2 =$ ____	$10 - 6 - 2 =$ ____
$8 - 1 - 3 =$ ____	$9 - 2 - 5 =$ ____	$10 - 1 - 4 =$ ____

2. Solve. You can draw pictures to help.

a. Mary had ten cookies. She gave two
   to her brother and two to her sister.
   How many does she have left?

b. Seven birds were in the tree. Three flew away.
   After a while, one more flew away.
   How many birds were left in the tree?

c. Eight cars were in the parking lot. Then
   three cars left. After that two more cars
   left. How many cars are there now?

d. Jack had $5. His mom gave him $1,
   and his dad gave him $2.
   How many dollars does Jack have now?

You can subtract two numbers one at a time:

$$8 - 2 - 3$$
$$\diagdown \diagup$$
$$6 \quad - 3 = 3$$

First take away 2. That leaves 6. Then, from 6, subtract 3. That leaves 3.

OR you can subtract their total:

$$8 - 2 - 3$$
$$\diagdown \diagup$$
$$8 - 5 \quad = 3$$

Check how much you need to subtract or take away *in total*. You need to subtract 2 and 3 — a total of 5. So subtract $8 - 5 = 3$.

3. Subtract by either method.

a.	b.	c.
$7 - 2 - 3 =$ ____	$9 - 7 - 1 =$ ____	$7 - 5 - 1 =$ ____
$9 - 2 - 6 =$ ____	$6 - 2 - 2 =$ ____	$10 - 6 - 1 =$ ____

4. Solve. Compare the two problems and their results.

a.	b.	c.
$10 - 3 - 2 =$ ____	$7 - 3 - 3 =$ ____	$9 - 6 - 1 =$ ____
$10 - 3 - 3 =$ ____	$7 - 4 - 3 =$ ____	$8 - 6 - 1 =$ ____

5. Match the subtraction problems to the correct pictures.

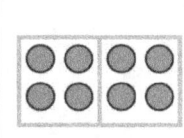

a. $8 - 2 - 2 - 2 - 2 = 0$

b. $8 - 4 - 4 = 0$

c. $6 - 2 - 2 - 2 = 0$

d. $6 - 3 - 3 = 0$

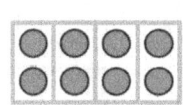

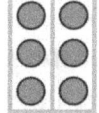

Puzzle Corner Here are some problems with four numbers!

$9 - 3 - 2 - 1 =$ _____ $10 - 1 - 2 - 1 =$ _____ $8 - 4 - 1 - 2 =$ _____

Review — Facts with 6, 7, and 8

1. Practice addition and subtraction facts with 6, 7, and 8.

a.	b.	c.	d.
$0 + \underline{\hspace{1cm}} = 8$	$3 + \underline{\hspace{1cm}} = 7$	$6 - \underline{\hspace{1cm}} = 2$	$7 - \underline{\hspace{1cm}} = 2$
$3 + \underline{\hspace{1cm}} = 8$	$5 + \underline{\hspace{1cm}} = 7$	$6 - \underline{\hspace{1cm}} = 5$	$8 - \underline{\hspace{1cm}} = 3$
$2 + \underline{\hspace{1cm}} = 8$	$1 + \underline{\hspace{1cm}} = 7$	$6 - \underline{\hspace{1cm}} = 3$	$6 - \underline{\hspace{1cm}} = 1$
$6 + \underline{\hspace{1cm}} = 8$	$6 + \underline{\hspace{1cm}} = 7$	$6 - \underline{\hspace{1cm}} = 4$	$8 - \underline{\hspace{1cm}} = 4$
$5 + \underline{\hspace{1cm}} = 8$	$2 + \underline{\hspace{1cm}} = 7$	$6 - \underline{\hspace{1cm}} = 1$	$7 - \underline{\hspace{1cm}} = 4$

2. First add and subtract. Write the answers in the boxes below. Then compare, and write $<$, $>$ or $=$.

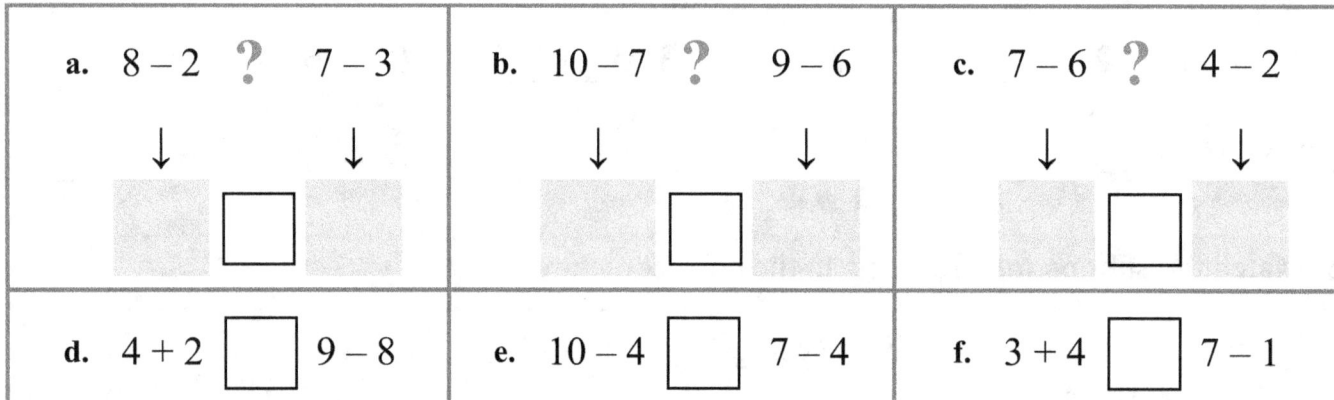

a. $8 - 2$ **?** $7 - 3$

b. $10 - 7$ **?** $9 - 6$

c. $7 - 6$ **?** $4 - 2$

d. $4 + 2 \ \square \ 9 - 8$

e. $10 - 4 \ \square \ 7 - 4$

f. $3 + 4 \ \square \ 7 - 1$

3. Solve.

a. Luisa and Caleb were playing a game. Luisa had 9 game pieces and Caleb had 4. How many more game pieces did Luisa have than Caleb?

b. Luisa gave one game piece to Caleb. Now who has more game pieces?

How many more?

4. Complete. Then draw lines to connect the facts from the same fact family.

_____ − 5 = 1	7 − 5 = _____	8 − 3 = _____
2 + _____ = 7	_____ + 2 = 6	5 + _____ = 7
8 − _____ = 3	6 − 1 = _____	1 + 5 = _____
_____ + 2 = 8	5 + _____ = 8	8 − 6 = _____
6 − 4 = _____	8 − _____ = 6	2 + 4 = _____

5. Complete. Then draw lines to connect the facts from the same fact family.

3 + _____ = 7	_____ + 6 = 7	8 − _____ = 7
6 − _____ = 3	_____ − 7 = 1	1 + 6 = _____
_____ + 1 = 8	3 + 3 = _____	3 + _____ = 6
_____ − 4 = 4	4 + _____ = 7	8 − 4 = _____
7 − 1 = _____	8 − _____ = 4	7 − _____ = 4

Puzzle Corner

What numbers can go into the squares? All the numbers are less than 10. Guess and check!

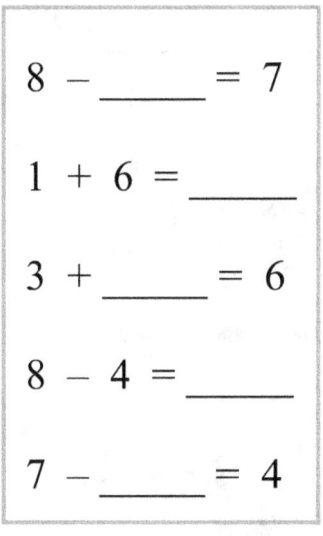

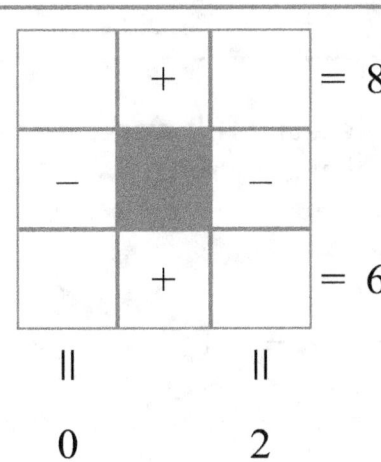

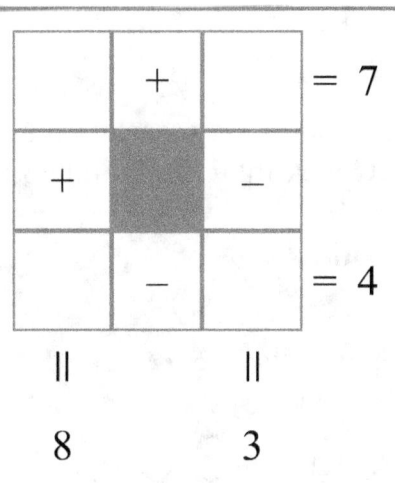

Review — Facts with 9 and 10

1. Practice addition and subtraction facts with 9 and 10.

a.	b.	c.	d.
$4 + \underline{\hspace{1cm}} = 9$	$5 + \underline{\hspace{1cm}} = 10$	$10 - \underline{\hspace{1cm}} = 1$	$9 - \underline{\hspace{1cm}} = 2$
$1 + \underline{\hspace{1cm}} = 9$	$2 + \underline{\hspace{1cm}} = 10$	$10 - \underline{\hspace{1cm}} = 7$	$9 - \underline{\hspace{1cm}} = 6$
$6 + \underline{\hspace{1cm}} = 9$	$3 + \underline{\hspace{1cm}} = 10$	$10 - \underline{\hspace{1cm}} = 5$	$9 - \underline{\hspace{1cm}} = 8$
$2 + \underline{\hspace{1cm}} = 9$	$4 + \underline{\hspace{1cm}} = 10$	$10 - \underline{\hspace{1cm}} = 8$	$9 - \underline{\hspace{1cm}} = 5$

2. Match the addition problems to the right pictures and solve them.

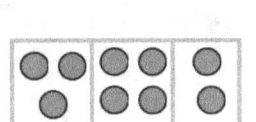

a. $2 + 3 + 3 = \underline{\hspace{2cm}}$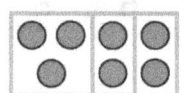

b. $3 + 2 + 2 = \underline{\hspace{2cm}}$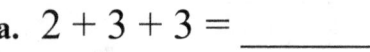

c. $1 + 2 + 2 = \underline{\hspace{2cm}}$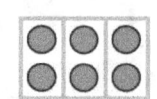

d. $3 + 4 + 2 = \underline{\hspace{2cm}}$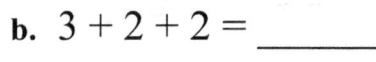

e. $3 + 3 + 3 = \underline{\hspace{2cm}}$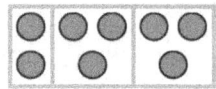

f. $2 + 2 + 2 = \underline{\hspace{2cm}}$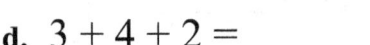

3. Time to play teacher again! Ann, Joe, and Bill worked some math problems. Check their work and correct any that are wrong.

Ann:	Joe:	Bill:
a. $5 - 0 = \triangle 5$	c. $9 - 4 = \triangle 6$	e. $7 - \triangle 5 = 3$
b. $10 - \triangle 3 = 6$	d. $6 - \triangle 4 = 2$	f. $\triangle 8 - 6 = 2$

4. Fill in the missing numbers. Draw lines to connect the facts that belong to the same fact family.

$9 - \rule{2em}{0.4pt} = 7$	$\rule{2em}{0.4pt} + 2 = 9$	$9 - \rule{2em}{0.4pt} = 5$
$9 - \rule{2em}{0.4pt} = 6$	$8 + \rule{2em}{0.4pt} = 9$	$9 - 6 = \rule{2em}{0.4pt}$
$9 - 1 = \rule{2em}{0.4pt}$	$\rule{2em}{0.4pt} + 5 = 9$	$9 - \rule{2em}{0.4pt} = 2$
$9 - \rule{2em}{0.4pt} = 4$	$3 + \rule{2em}{0.4pt} = 9$	$\rule{2em}{0.4pt} + 8 = 9$

5. **a.** Draw a line to connect each pair of numbers that add up to 9. Which number is left by itself?

```
0        7      2
    2       8      5
1       6
   9        4    3    4
      5
    1            3    8
7            6            9
```

b. Draw a line to connect each pair of numbers that add up to 10. Which number is left by itself?

```
3            7    10
     2          8       9
1        6
    9         4    0    2
       5
    1          3    8
7            6            5
```

6. Fill in the missing numbers. Draw lines to connect the facts that belong to the same fact family.

$10 - \rule{2em}{0.4pt} = 8$	$\rule{2em}{0.4pt} + 9 = 10$	$\rule{2em}{0.4pt} + 1 = 10$
$10 - \rule{2em}{0.4pt} = 5$	$4 + \rule{2em}{0.4pt} = 10$	$10 - 5 = \rule{2em}{0.4pt}$
$10 - \rule{2em}{0.4pt} = 1$	$5 + \rule{2em}{0.4pt} = 10$	$10 - 4 = \rule{2em}{0.4pt}$
$10 - 3 = \rule{2em}{0.4pt}$	$2 + \rule{2em}{0.4pt} = 10$	$\rule{2em}{0.4pt} + 3 = 10$
$10 - 6 = \rule{2em}{0.4pt}$	$\rule{2em}{0.4pt} + 7 = 10$	$10 - \rule{2em}{0.4pt} = 8$

7. Solve.

a. Millie has two boxes of crayons. Ken has seven boxes.
How many more boxes does Ken have?

b. Mike counted his toy cars. He has three yellow ones, four blue ones,
and three red ones. How many toy cars does he have in all?

c. There were four birds in a tree. Four more flew in.
How many birds are there now?

Look! Five of them just flew away!
How many birds are there now?

d. Elisa knows she has ten crayons. She can only find four.
How many are missing?

e. A ten-piece puzzle has two pieces missing.
How many pieces are there now?

Chapter 5: Time
Introduction

The fifth chapter covers reading an analog clock to whole hours and to half hours, and some basics of time and the calendar.

In the first lesson, we use an analog clock that only has the hour hand. We omit the minute hand for a reason: this way the child can concentrate on the hour hand only, and learning to tell whole and half hours becomes much easier. We also practice telling what time it is one hour or a half-hour later than a given time.

The next lesson focuses on minutes. The aim of this lesson is to learn that one hour is 60 minutes, that a half-hour is 30 minutes, and how the phrases "o'clock" and "half past" relate to the hours and minutes. For example, the child is to learn that "half past eight" is written 8:30, and the "30" part is the number of minutes, so half an hour is just 30 minutes.

This lesson also includes a few exercises about reading the clock to five-minute intervals using a special clock that includes the numbers for the minute hand; however, these can be skipped if desired, because in second grade, the student will get a lot of practice reading the clock to the nearest five minutes.

I have included one lesson about time order. The topics in this lesson are hopefully already familiar to the student.

The next lesson deals with morning and afternoon hours: AM and PM. The goal is for the student to understand that the clock starts at 12 midnight, and goes through all the A.M. hours from 1 to 12 until it is 12 noon, and then goes through all of the P.M. hours from 1 to 12 until it is 12 midnight again.

We will also briefly look at the calendar and practice the names of the months.

Reading the clock is a skill that can and should be practiced in everyday situations from now on so that children can learn by experience, and not just by filling in pages in their math book.

Pacing Suggestion for Chapter 5

Please add one day to the pacing for the test if you will use it. Note that the specific lessons in the chapter can take several days to finish. They are not "daily lessons." As a general guideline, first graders should finish 1-2 pages daily or 7-9 pages a week. Please also see the user guide at https://www.mathmammoth.com/userguides/ .

The Lessons in Chapter 5	page	span	suggested pacing	your pacing
Whole and Half Hours	38	*4 pages*	2 days	
Minutes and Half Hours.................................	42	*4 pages*	2 days	
Time Order	46	*2 pages*	1 day	
AM and PM	48	*3 pages*	2 days	
The Calendar	51	*2 pages*	2 days	
Review - Half Hours	53	*1 page*	1 day	
Chapter 5 Test (optional)				
TOTALS		*16 pages*	10 days	

Games and Activities

Tell the Time!

You need: An analog clock that allows you to turn the clock hands, or an app that allows you to do so.

In this activity, ask your child or student to turn the clock hands to a specific time (whole hours or half hours). Once they do so, then it is their turn to give you a time that you will set the clock to. You can use random times, and also important, specific times, such as, "Set the clock to the time when we eat supper (6 o'clock)."

Find the Date!

You need: A wall calendar

In this simple activity, ask your child or student to find a specific date on the calendar. Once they do so, then it is their turn to tell you a date to find. You can find random dates, and also important, specific dates, such as, "Find my birthday, September 12."

Months Match

This is a simple activity to practice matching the names of the months to their numbers.

You need: A set of 12 number cards with numbers from 1 to 12 on them. You can use cards from a standard deck if your child understands Jack as 11 and Queen as 12.

Shuffle the cards. Ask the child to turn the cards from the deck one by one, and at each card, say the name of the month that corresponds to that number. For example, if the child gets 7, they should say "July".

Once the child can go through all 12 cards without any mistakes, give them a small reward.

Games and Activities at Math Mammoth Practice Zone

Telling Time
Practice telling time on an analog clock with this interactive online exercise. Choose "Whole hours" or "Whole hours and half hours" for this grade level.
https://www.mathmammoth.com/practice/tell-time

Further Resources on the Internet

We have compiled a list of Internet resources that match the topics in this chapter, including pages that offer:

- **online practice** for concepts;
- online **games**, or occasionally, printable games;
- **animations** and interactive **illustrations** of math concepts;
- **articles** that teach a math concept.

We heartily recommend you take a look! Many of our customers love using these resources to supplement the bookwork. You can use these resources as you see fit for extra practice, to illustrate a concept better and even just for some fun. Enjoy!

https://l.mathmammoth.com/gr1ch5

Scan me

Whole and Half Hours

In this lesson, the clock only has one hand – the HOUR hand.

The hour hand points to four –
it is four hours, or "four o'clock."

The hour hand points to eleven –
it is eleven hours, or "eleven o'clock".

The hour hand moves slowly around the clock
face: from 1 to 2 to 3, and so on.

When the hour hand moves from 1 to 2, exactly
one hour of time has passed.

The same is true when the hour hand moves from
2 to 3. It takes the hour-hand one hour to do that.

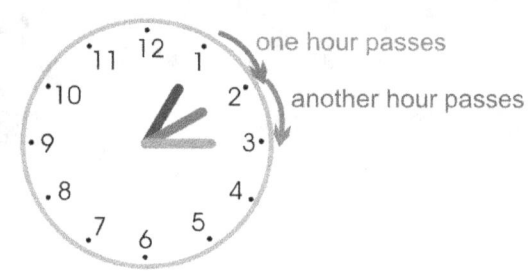

On this clock, the hour hand pointed to 5
when it was five o'clock.

Then it moved to **halfway** between 5 and 6.
We say it is **half past five**.

It takes the hour hand one-half hour to move from five
to halfway between five and six.

Here the hour hand has moved past eight o'clock, and is halfway
between 8 and 9. We say it is half past eight.

In half an hour, it will be nine o'clock.

1. Write the time under each clock face.

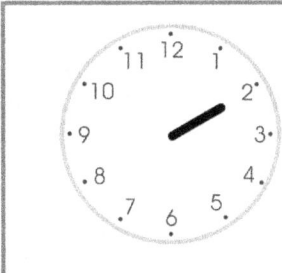   

a. _____ o'clock   **b.** _____ o'clock   **c.** _____ o'clock   **d.** _____ o'clock

2. Draw the hour hand.

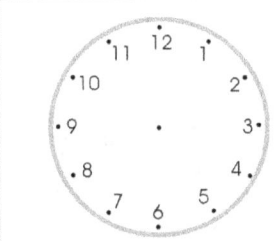 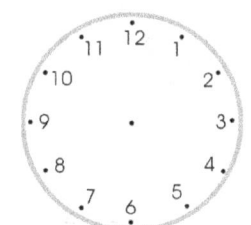 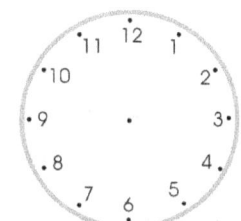 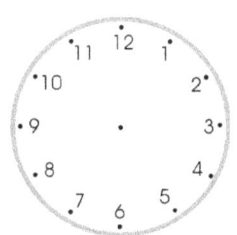

a. Five o'clock   **b.** Eight o'clock   **c.** Twelve o'clock   **d.** Seven o'clock

3. Write the time.

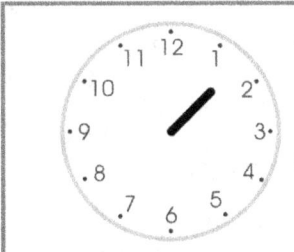 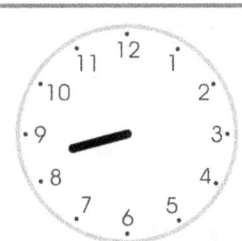  

a. Half past _____   **b.** Half past _____   **c.** Half past _____   **d.** Half past _____

4. Draw the hour hand.

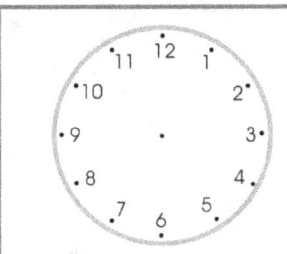 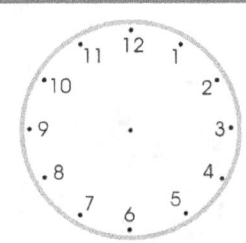 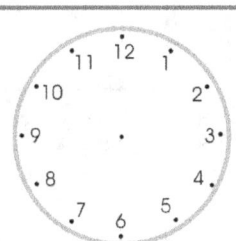 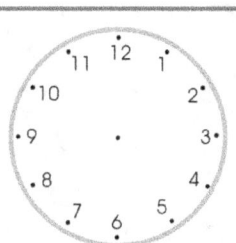

a. Half past six   **b.** Half past three   **c.** Half past two   **d.** Half past four

5. Write the time!

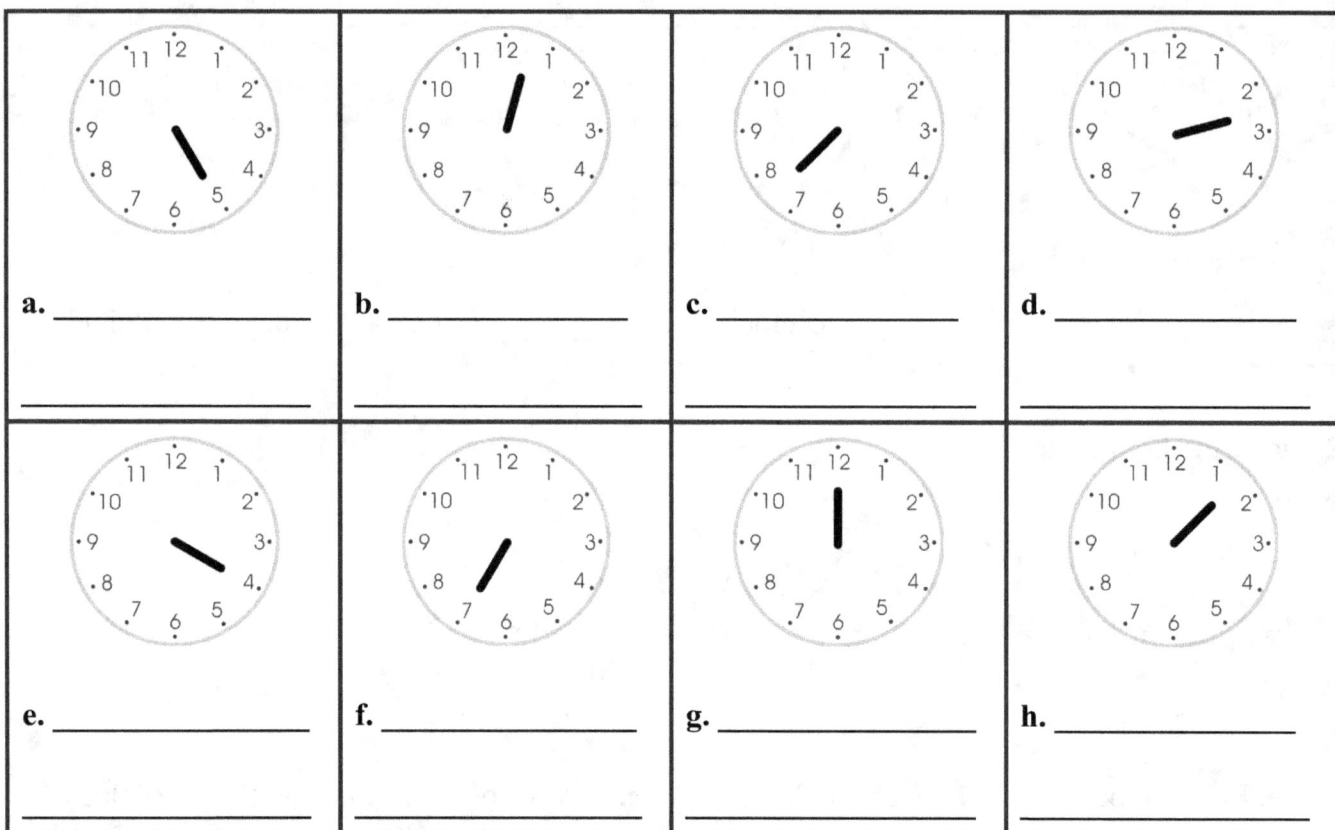

a. _____

b. _____

c. _____

d. _____

_____

_____

_____

_____

e. _____

f. _____

g. _____

h. _____

_____

_____

_____

_____

6. Draw an hour hand on each clock. In the second row, show the time a half-hour later. In the third row, show the time another half-hour later than the clock in the second row.

Draw the hour hand.	a. Five o'clock	b. One o'clock	c. Half past six	d. Half past three
A half-hour later →				
Another half-hour later →				

7. Draw the hour hand on each clock. Write the time that the clock will show a half-hour later.

	a. Three o'clock	**b.** Eleven o'clock	**c.** Half past five	**d.** Half past eleven
1/2 hour later →	half past _____	half past _____	_____ o'clock	_____ o'clock

8. Write the time that the clock shows. Then write what the time will be an hour later.

	a. _____ o'clock	**b.** _____ o'clock	**c.** half past _____	**d.** half past _____
An hour later →				

9. Draw the hour hand on the clock face. Write what it will be an hour later.

	a. Three o'clock	**b.** Eleven o'clock	**c.** Half-past five	**d.** Half past eleven
An hour later →				

Minutes and Half Hours

The minute hand on the clock is the thinner
and longer hand.

It shows us the minutes - but the numbers 1-12
on the clock face do NOT tell us the minutes.

The green numbers are for the minute hand. They
are not normally written on the clock face at all.

The time on the clock is 4 o'clock, or 4:00,
which is four hours and zero minutes.

 Find a clock that has a knob you can turn to move the hour and minute hands. Set the hour hand pointing to one, and the minute hand pointing straight "up".

Move the hour hand from 1 to 2, and observe the minute hand!

Now move the hour hand from 2 to 3.
What does the minute hand do?

_____

If you make the hour hand travel from 7 to 10,
how many "rounds" does the minute-hand make? _____

Make the minute hand travel backwards as well.

It is 1 o'clock
(and 0 minutes)

1. You know these from the previous lesson. Now the minute hand is added.
   Write the time using the expressions *o'clock* or *half past*.

a. _____

_____

b. _____

_____

c. _____

_____

d. _____

_____

1 HOUR = 60 MINUTES!

Use your clock. How much time passes when the hour hand travels from 1 to 2? _____ hour

At the same time, the minute hand travels from 0 to 60 minutes, or once around the clock.

It is 4:00.

Now let the hour hand travel from 2 to half-past 2 (only half an hour). The minute hand traveled from 0 minutes to 30 minutes.

We write the time as 2:30.
That means 2 hours and 30 minutes.

The green numbers are for the minutes. Just count by fives to learn them!

Then let the hour hand travel from half-past 2 to 3 — another half an hour.

The minute hand traveled from 30 minutes to _____ (or 0) minutes.

1 hour = 60 minutes.

The hour hand is past 3, and the minute hand points to 20. The time is 3:20, which is 3 hours and 20 minutes.

The hour hand *looks like* pointing to eight, but it's not quite eight o'clock. The hour hand is past seven and not yet on eight, and that is why we say the time is "seven forty," or 7:40, which is 7 hours and 40 minutes.

Again, the hour hand looks like it's pointing to six, but it's not yet six o'clock (though it is almost six). The hour hand has passed five, so we say it is "five fifty-five," or 5:55, which is 5 hours and 55 minutes.

2. Write the time in two ways: (1) using the expressions *o'clock* or *half past* and (2) with numbers.

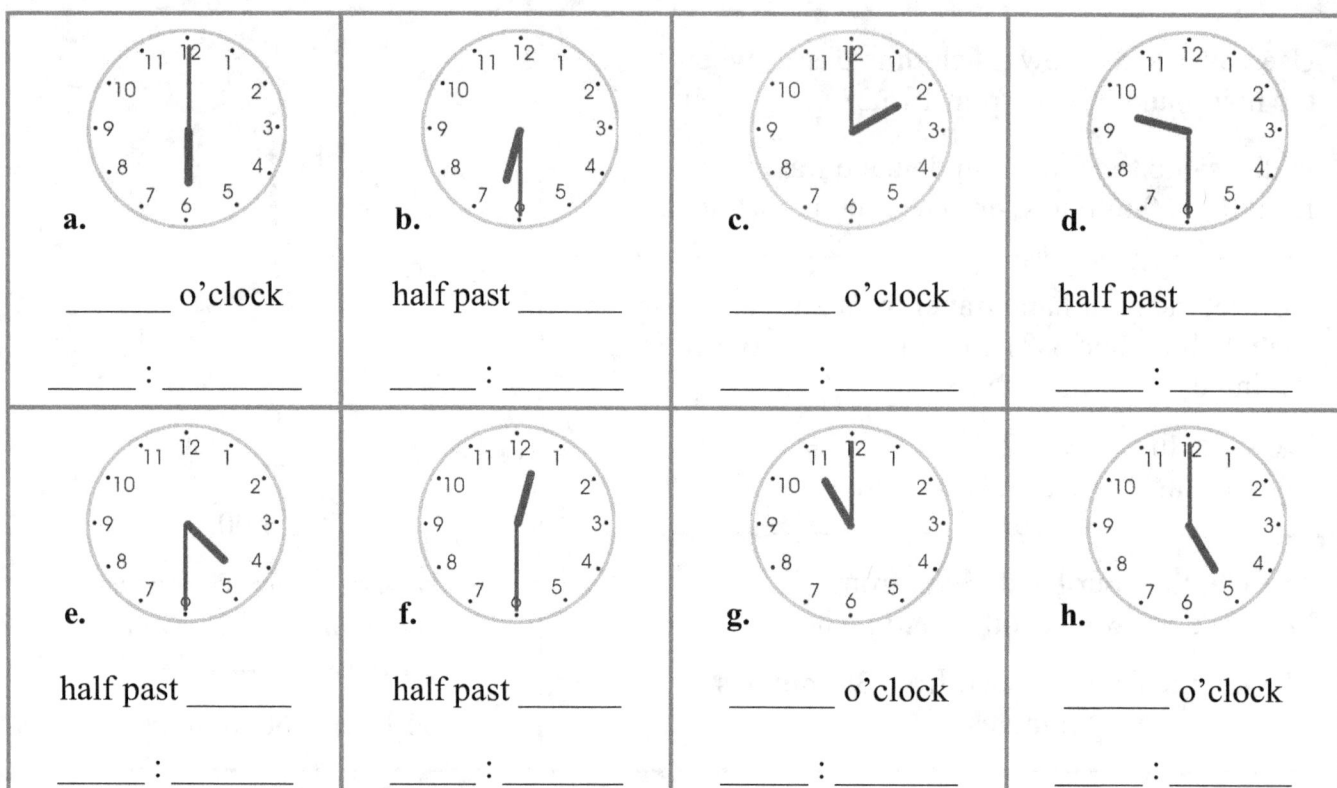

a. _____ o'clock

____ : _____

b. half past _____

____ : _____

c. _____ o'clock

____ : _____

d. half past _____

____ : _____

e. half past _____

____ : _____

f. half past _____

____ : _____

g. _____ o'clock

____ : _____

h. _____ o'clock

____ : _____

3. Write the time a half-hour later and another half-hour later. Use numbers.

Now it is:	a. 3:00	b. 6:30	c. 9:00	d. 2:30	e. 11:30
1/2 hour later it is:					
another 1/2 hour later:					

4. Write the time an hour later. Use numbers.

Now it is:	a. 2:00	b. 5:30	c. 11:30	d. 12:00	e. 8:30
an hour later it is:					

5. How many hours pass? Turn the hands on your practice clock.

a. From 5:00 to 8:00 _____ hour(s)   b. From 5:00 to 6:00 _____ hour(s)

c. From 8:00 to 8:30 _____ hours(s)   d. From 9:00 to 11:00 _____ hour(s)

e. From 4:00 to 9:00 _____ hour(s)   f. From 11:30 to 12:00 _____ hour(s)

6. Write the time using the special clock.

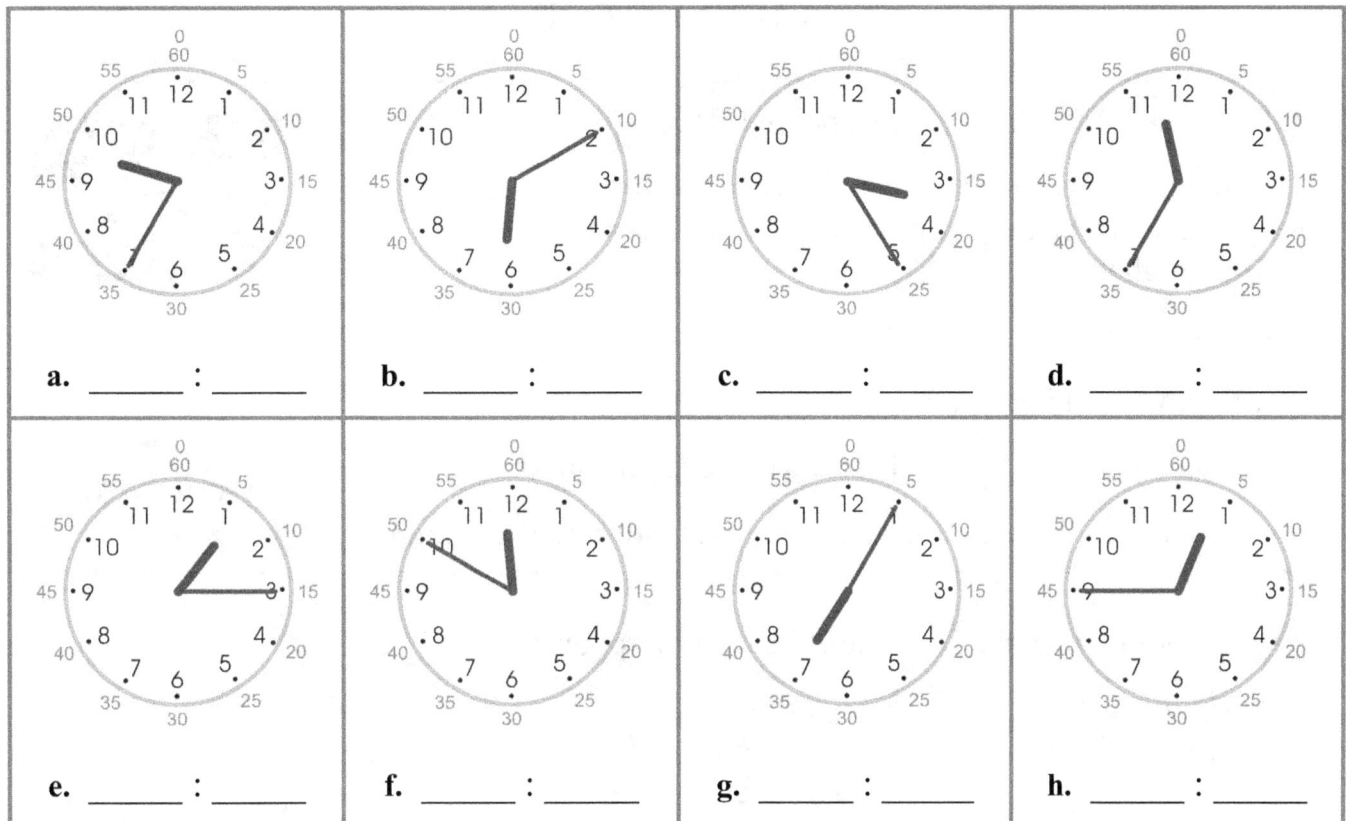

a. _____ : _____

b. _____ : _____

c. _____ : _____

d. _____ : _____

e. _____ : _____

f. _____ : _____

g. _____ : _____

h. _____ : _____

7. Write the time. Don't worry if it is difficult. You will practice this more in second grade.

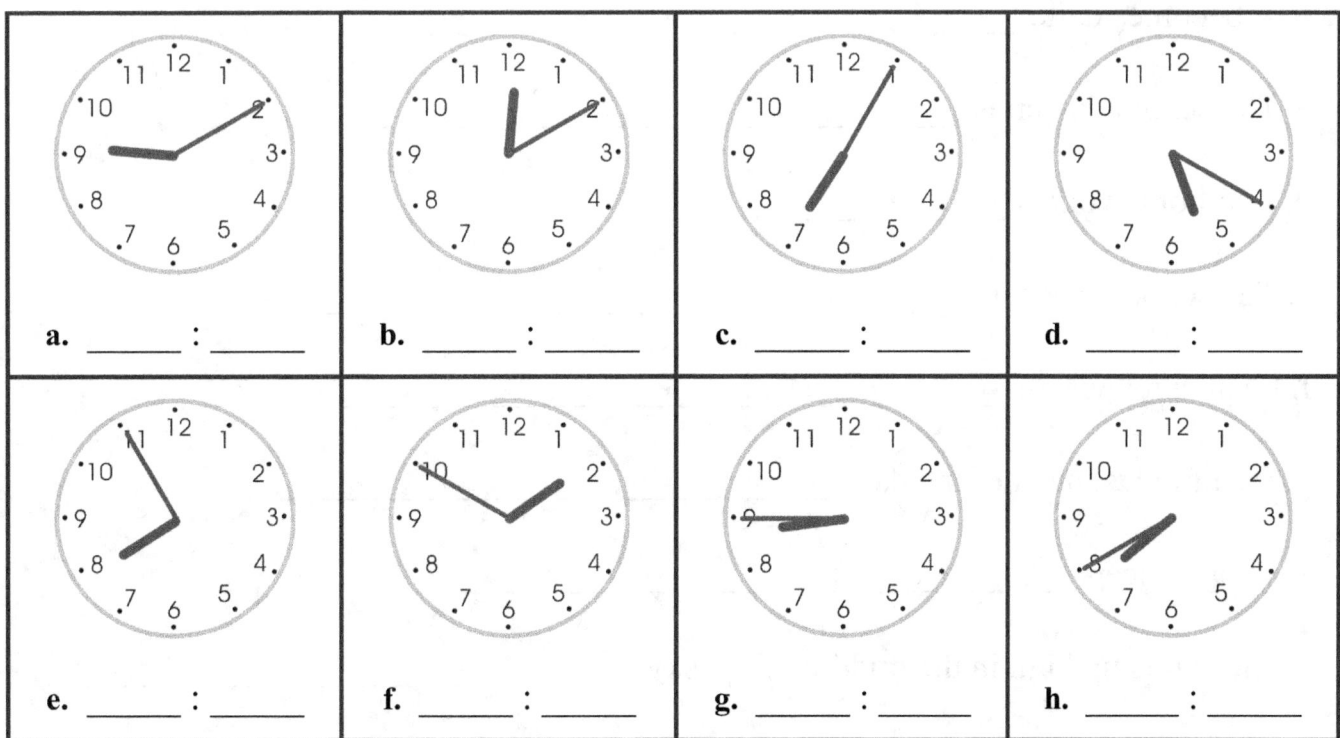

a. _____ : _____

b. _____ : _____

c. _____ : _____

d. _____ : _____

e. _____ : _____

f. _____ : _____

g. _____ : _____

h. _____ : _____

Time Order

Night is when it is dark and most people sleep.

After night, comes morning. That is when the sun rises and the day starts. All morning, the sun keeps climbing higher.

Noon is when the sun is at its highest point in the sky.

After that, we call it afternoon.

Afternoon ends at 6 o'clock. That's when evening starts.

The sun sets and the darkness starts some time during the evening.

1. Here are some events. Tell when the event happens:

- early morning
- morning
- noon

- afternoon
- evening
- night

a. Birds start singing and the sun rises. _____

b. It becomes dark. _____

c. The school day ends. _____

d. I do homework. _____

e. The school day starts. _____

f. I watch television. _____

g. I eat my last meal of the day. _____

h. People sleep. _____

i. The sun is up high in the middle of the sky. _____

j. I do my chores. _____

2. Add the words *yesterday*, *today*, and *tomorrow*. The events are not in order.

a. _____, Paul studied for the test.

_____, Paul will know the results.

_____, Paul has a math test.

b. _____, Mom makes a birthday cake!

_____, we eat the leftovers.

_____, Mom bought eggs, flour, sugar, and fruit.

3. Put the events in order. What happens first? Next? Last? Mark 1, 2, and 3.

Jane's ankle is hurt. 	Colored pencils! 	John goes to school. 
Jane rides her bike. 	Rick gets a gift. 	In the afternoon, John plays soccer. 
Jane crashes into a tree. 	Rick draws an airplane. 	John likes math class. 

4. Circle the event that takes a longer time.

a. Prepare a meal.	Eat a banana.	**b.** Read a book.	Read a letter.
c. Take a shower.	Clean the house.	**d.** A math class	A soccer game
e. Dress a baby.	Go shopping for food.	**f.** Play a game of cards	Brush your teeth

AM and PM

The 12 clock hours before noon are marked with AM.
The 12 clock hours after noon are marked with PM.

AM is short for Latin *Ante Meridiem*, which means "before noon."
PM comes from Latin *Post Meridiem*, which means "after noon."

NOON is when the sun is at its highest point. It is the middle of the day, and people usually eat lunch around noon.

$$12-1-2-3-4-5-6-7-8-9-10-11-12-1-2-3-4-5-6-7-8-9-10-11-12$$

Midnight   AM   Noon   PM   Midnight

So, from midnight, for the rest of the night, at sunrise, and all through the morning until NOON (when the sun is at its highest), the hours are AM.

After noon, in the afternoon, when the sun is going down, when the sun sets, and all through the evening until midnight, the hours are PM.

The clock has only 12 hours, but our day and night have 24 hours all total.

That is why the hour hand travels through those 12 hours *two* times in one day-night period: 12 + 12 = 24.

So every clock hour (such as 7:00) happens two times during one day-night period.

7:00 AM
This is before noon.

7:00 PM
This is after noon, during the evening.

Because of the meaning of AM and PM, noon is technically neither. The same is true of midnight. However, most people consider noon to be 12 PM (the beginning of the PM) and midnight to be 12 AM (the beginning of the AM). You can say, "12 o'clock noon" or "12 o'clock midnight" to avoid confusion.

1. **a.** Mark on the hour line when you usually wake up and when you usually go to bed.

   b. Mark your sleeping hours with black.

   c. Mark on the hour line when you eat some of your main meals.

$$12-1-2-3-4-5-6-7-8-9-10-11-12-1-2-3-4-5-6-7-8-9-10-11-12$$

Midnight   AM   Noon   PM   Midnight

2. Mark on the hour line when the SUN usually rises and when it usually sets. Color the nighttime hours black.

(Ask your teacher to help you make different charts for different seasons.)

a. the middle of winter

12 – 1 – 2 – 3 – 4 – 5 – 6 – 7 – 8 – 9 – 10 – 11 – 12 – 1 – 2 – 3 – 4 – 5 – 6 – 7 – 8 – 9 – 10 – 11 – 12

Midnight   AM   Noon   PM   Midnight

b. spring

12 – 1 – 2 – 3 – 4 – 5 – 6 – 7 – 8 – 9 – 10 – 11 – 12 – 1 – 2 – 3 – 4 – 5 – 6 – 7 – 8 – 9 – 10 – 11 – 12

Midnight   AM   Noon   PM   Midnight

c. summer

12 – 1 – 2 – 3 – 4 – 5 – 6 – 7 – 8 – 9 – 10 – 11 – 12 – 1 – 2 – 3 – 4 – 5 – 6 – 7 – 8 – 9 – 10 – 11 – 12

Midnight   AM   Noon   PM   Midnight

3. Write what you are usually doing at these times of day.

a. 5:00 AM _____

b. 8:00 AM _____

c. 12:00 noon _____

d. 3:00 PM _____

e. 6:00 PM _____

f. 9:00 PM _____

g. 12:00 midnight _____

4. Is it before noon, or not? Write either AM or PM to each situation.

a. The birds are awake and singing, but most people are still in bed.	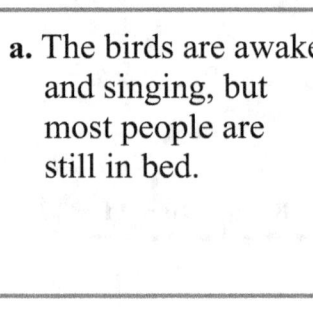	**b.** Mom is watching the evening news on TV.	
c. It is time to start schoolwork!		**d.** The sun is shining brightly.	
e. It is quiet and dark outside.		**f.** It is night.	
g. We are eating lunch.		**h.** Jack went to bed two hours ago.	

5. Draw the hour and minute hands on the clock so it matches the situation.
   Then write the time, including AM or PM.

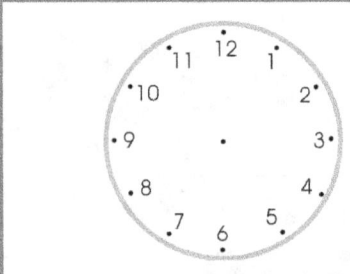

a. I wake up

_____ : _____ _____

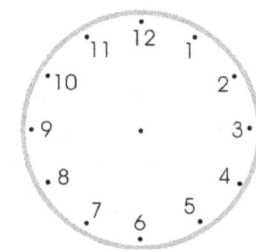

b. I eat lunch

_____ : _____ _____

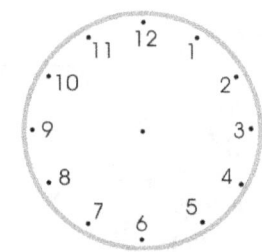

c. I eat supper

_____ : _____ _____

The Calendar

January						
Su	Mo	Tu	We	Th	Fr	Sa
				1	2	3
4	5	6	7	8	9	10
11	12	13	14	15	16	17
18	19	20	21	22	23	24
25	26	27	28	29	30	31

February						
Su	Mo	Tu	We	Th	Fr	Sa
1	2	3	4	5	6	7
8	9	10	11	12	13	14
15	16	17	18	19	20	21
22	23	24	25	26	27	28

March						
Su	Mo	Tu	We	Th	Fr	Sa
1	2	3	4	5	6	7
8	9	10	11	12	13	14
15	16	17	18	19	20	21
22	23	24	25	26	27	28
29	30	31				

April						
Su	Mo	Tu	We	Th	Fr	Sa
			1	2	3	4
5	6	7	8	9	10	11
12	13	14	15	16	17	18
19	20	21	22	23	24	25
26	27	28	29	30		

May						
Su	Mo	Tu	We	Th	Fr	Sa
					1	2
3	4	5	6	7	8	9
10	11	12	13	14	15	16
17	18	19	20	21	22	23
24	25	26	27	28	29	30
31						

June						
Su	Mo	Tu	We	Th	Fr	Sa
	1	2	3	4	5	6
7	8	9	10	11	12	13
14	15	16	17	18	19	20
21	22	23	24	25	26	27
28	29	30				

July						
Su	Mo	Tu	We	Th	Fr	Sa
			1	2	3	4
5	6	7	8	9	10	11
12	13	14	15	16	17	18
19	20	21	22	23	24	25
26	27	28	29	30	31	

August						
Su	Mo	Tu	We	Th	Fr	Sa
						1
2	3	4	5	6	7	8
9	10	11	12	13	14	15
16	17	18	19	20	21	22
23	24	25	26	27	28	29
30	31					

September						
Su	Mo	Tu	We	Th	Fr	Sa
		1	2	3	4	5
6	7	8	9	10	11	12
13	14	15	16	17	18	19
20	21	22	23	24	25	26
27	28	29	30			

October						
Su	Mo	Tu	We	Th	Fr	Sa
				1	2	3
4	5	6	7	8	9	10
11	12	13	14	15	16	17
18	19	20	21	22	23	24
25	26	27	28	29	30	31

November						
Su	Mo	Tu	We	Th	Fr	Sa
1	2	3	4	5	6	7
8	9	10	11	12	13	14
15	16	17	18	19	20	21
22	23	24	25	26	27	28
29	30					

December						
Su	Mo	Tu	We	Th	Fr	Sa
		1	2	3	4	5
6	7	8	9	10	11	12
13	14	15	16	17	18	19
20	21	22	23	24	25	26
27	28	29	30	31		

1. Circle these dates on the calendar above.

a. March 15

b. November 18

c. December 29

d. February 8

e. November 11

f. October 30

g. today

h. your birthday

i. any other important dates

2. What do the letters "Su Mo Tu We Th Fr Sa" mean?

3. Try not to look at the calendar. Write the name of the month that goes in between.

   a. August   _____   October

   b. March   _____   May

   c. June   _____   August

   d. January   _____   March

   e. October   _____   December

4. (Use the calendar above.) What day of the week is...

   a. January 4 _____   **b.** May 14   _____

   c. July 23 _____   **d.** March 10   _____

   e. today _____   **f.** your birthday _____

5. Find all of the MONDAYS in June. Write their dates below:

June _____	June _____	June _____	June _____	June _____

6. Find all of the THURSDAYS in April. Write their dates below:

April _____	April _____	April _____	April _____	April _____

7. Can you name the months in order from memory?
   Practice memorizing them in groups of four.

   • January, February, March, April;

   • May, June, July, August;

   • September, October, November, December.
     (These are all the months ending in "ber"!

Review - Half Hours

1. Write the time using the expressions *o'clock* or *half past*.

a. _____   b. _____   c. _____   d. _____

_____   _____   _____   _____

2. Write the time using numbers.

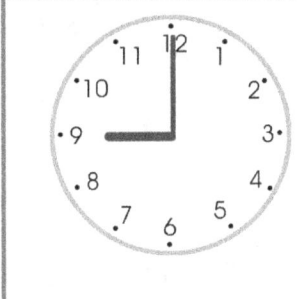   

a. _____ : _____   b. _____ : _____   c. _____ : _____   d. _____ : _____

3. Write the time at a half-hour and an hour later. Use numbers.

Now it is:	a. 5:00	b. 10:30	c. 12:00	d. 1:30	e. 5:30
A half-hour later, it is:					
An hour later, it is:					

4. Fill in either "AM" or "PM."

a. Jack wakes up. It is 8 _____.	b. Jack plays in the afternoon. It is 3 _____.
c. Jack is sleeping. It is dark. It is 2 _____.	d. Time for an evening snack! It is 8 _____.

Chapter 6: Shapes and Measuring
Introduction

This sixth chapter of *Math Mammoth Grade 1* covers basic shapes, halves and fourths, the concept of measuring, and measuring length with whole inches and centimeters.

The goals of this section are:

- The child composes and decomposes geometric figures (e.g., putting two triangles together to make a quadrilateral, or dividing a square into two triangles), and thereby builds understanding of part-whole relationships as well as the properties of the original and composite shapes.

- The child can draw lines with a ruler.

- The child understands that measuring length is a process of iterating (repeating) the unit of measure.

The lessons in this chapter can seem quite easy, but they are laying a proper foundation for geometric understanding in later years. For example, dividing shapes into parts not only makes the child familiar with the properties of the original shape and of its parts, but also helps to build an understanding of part-whole relationships for the study of fractions.

Don't forget to also check out the videos at https://www.mathmammoth.com/videos/ .

Note: If you have the electronic version of this book (a PDF file), you need to print the pages at 100%, instead of using "shrink to fit," "print to fit," or similar options from your printer. Printing with "shrink to fit" will cause the images to be slightly smaller than intended, and thus some exercises about measuring in inches and centimeters will no longer be a whole-number amount of inches or centimeters.

Pacing Suggestion for Chapter 6

Please add one day to the pacing for the test if you will use it. Note that the specific lessons in the chapter can take several days to finish. They are not "daily lessons." As a general guideline, first graders should finish 1-2 pages daily or 7-9 pages a week.

The Lessons in Chapter 6	page	span	suggested pacing	your pacing
Basic Shapes	57	*3 pages*	2 days	
Playing with Shapes	60	*2 pages*	1 day	
Drawing Basic Shapes	63	*3 pages*	2 days	
Practicing Basic Shapes and Patterns	66	*3 pages*	2 days	
Halves and Quarters	69	*4 pages*	2 days	
Measuring Length	73	*4 pages*	2 days	
Exploring Measuring	77	*2 pages*	1 day	
Measuring Lines in Inches	79	*3 pages*	2 days	
Measuring Lines in Centimeters	82	*2 pages*	1 day	
Three-Dimensional Shapes	84	*2 pages*	1 day	
Review Chapter 6	86	*1 page*	1 day	
Chapter 6 Test (optional)				
TOTALS		*29 pages*	17 days	

Games and Activities

Free Drawing

You need: paper, pencil, ruler

Ask the child to draw a certain number of dots on the paper, such as four or five, and then connect those with line segments (using a ruler!) to get a shape. Then the child can divide this shape into smaller shapes by drawing more lines. What shapes do they get? Encourage the child to experiment freely with such drawings.

Pattern Blocks/Tangram

A set of pattern blocks or a tangram game, where children use shapes to make new composite shapes, is something all children love. The list of further Internet resources gives links to free online versions. The below links give examples of sets on Amazon, but there are many more available. Look for a set that comes with pattern cards.

Coogam Wooden Pattern Blocks Set 130PCS
https://www.amazon.com/dp/B07MYYK64R/?tag=mathmammoth-20

LOVESTOWN 230 Pcs Wooden Pattern Blocks
https://www.amazon.com/dp/B08C37JMP2/?tag=mathmammoth-20

Measuring in Inches and Centimeters

The lessons in this chapter contains hands-on measurement activities. Beyond those, you can encourage the child to freely use rulers and measuring tapes around the house or classroom to measure anything and everything.

- Ask the child to find items that are longer than 12 cm (5 inches) but less than 30 cm (12 inches)

- Ask the child to measure their own bed, their own room, and so on.

- Ask the child to measure how tall someone is, by having the person lie down on the floor.

Games and Activities at Math Mammoth Practice Zone

Sort the Shapes
Drag each shape to the correct box.
https://www.mathmammoth.com/practice/sorting-game#questions=6&sort-by=shape

Further Resources on the Internet

We have compiled a list of Internet resources that match the topics in this chapter, including pages that offer:

- **online practice** for concepts;
- online **games**, or occasionally, printable games;
- **animations** and interactive **illustrations** of math concepts;
- **articles** that teach a math concept.

We heartily recommend you take a look! Many of our customers love using these resources to supplement the bookwork. You can use these resources as you see fit for extra practice, to illustrate a concept better and even just for some fun. Enjoy!

https://l.mathmammoth.com/gr1ch6

Scan me

Basic Shapes

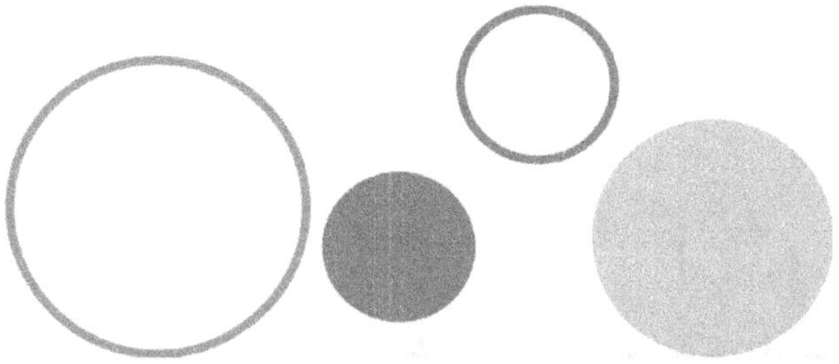

These are **circles**. They don't have any corners. They are perfectly round!

These are **triangles**. They have THREE corners and three sides.

These are **rectangles**. They have four "square" corners. They look like books!

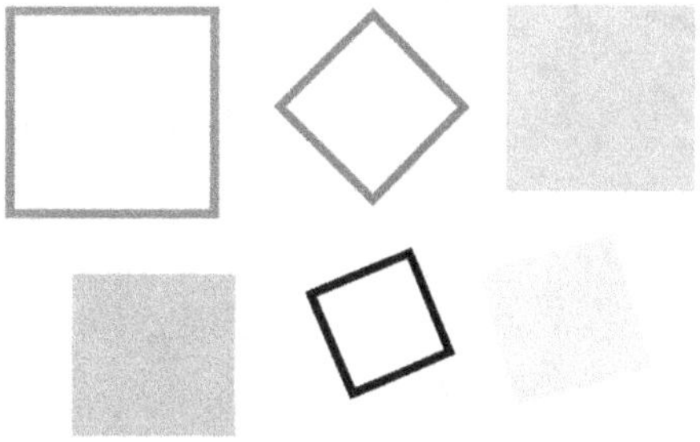

These are **squares**. Not only do squares have four square corners, but they also have four sides that are the same length.

1. Color the circles yellow; the squares red; the triangles green and the rectangles blue. One shape will not be colored.

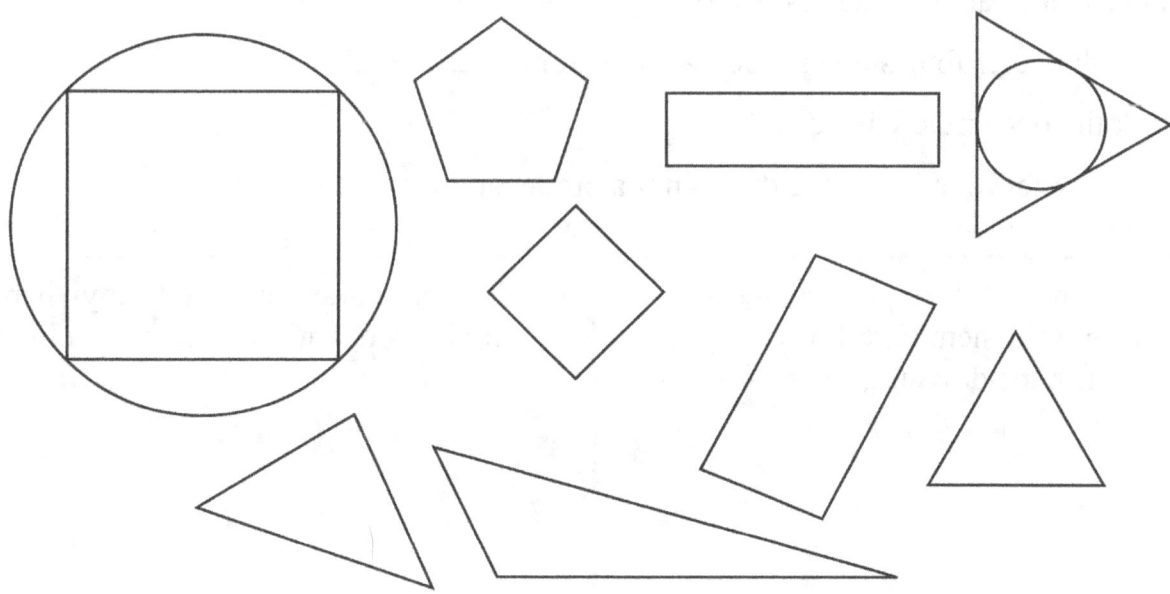

So what are these shapes?

They have four corners and four sides.
But they don't have four square corners,
like squares and rectangles do.

They are just **four-sided shapes** that are
not squares nor rectangles. In mathematics
we call them **quadrilaterals**.

"Quadri-" comes from *quattuor,* Latin for "four."
"Lateral" comes from *lateralis*, Latin for "side."

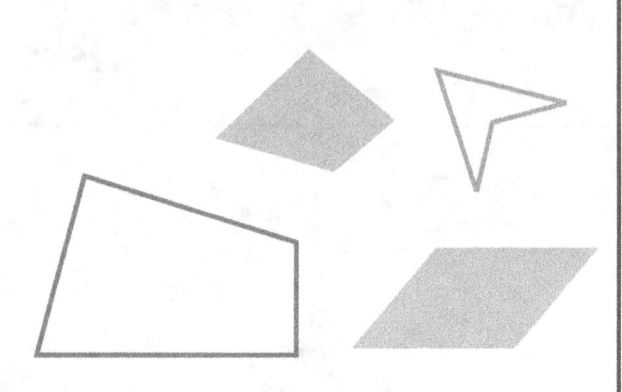

2. Write down how many corners each shape has.

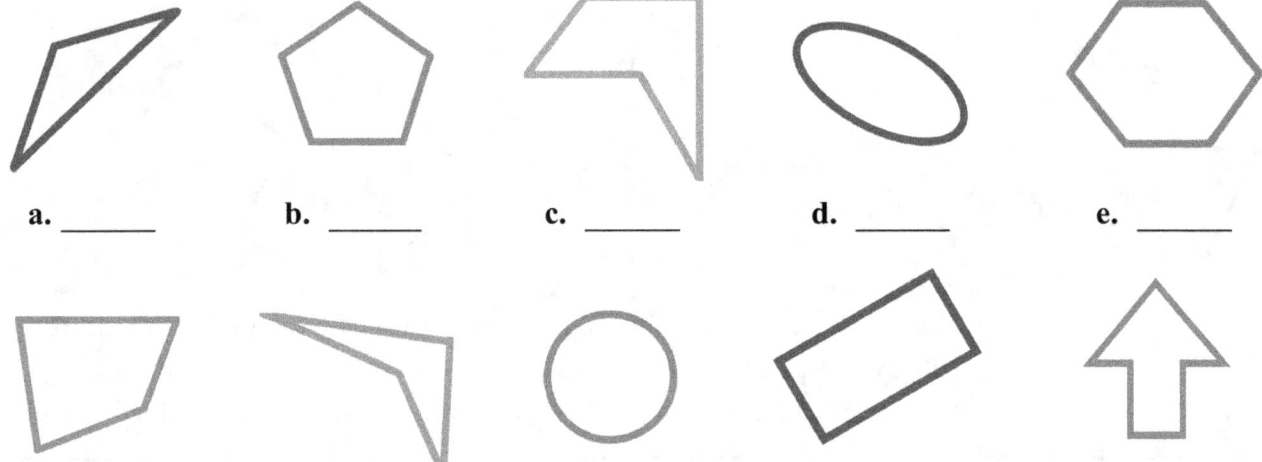

a. _____   b. _____   c. _____   d. _____   e. _____

f. _____   g. _____   h. _____   i. _____   j. _____

3. **a.** In the shapes above, there is one rectangle. Mark it with **R**.

   b. Mark the other four-sided shapes with **Q** (for quadrilateral).

   c. Mark the one circle with **C**.

   d. Find another rounded shape that is not a circle, and mark it with **E**.

4. **a.** Draw three dots anywhere in this space. Join them with lines. What shape do you get?	**b.** Once again draw three dots anywhere in this space, and join them with lines.

5. Draw a line from dot to dot so that you divide the shape into <u>two new shapes</u>. Use a ruler. How many sides do the new shapes have? How many corners?

a. The new shapes have _____ sides,

and _____ corners.

They are _____

b. The new shapes have _____ sides,

and _____ corners.

They are _____

c. The new shapes have _____ sides,

and _____ corners.

They are __*quadrilaterals*__

d. The new shapes have _____ sides,

and _____ corners.

They are _____

e. The new shapes have _____ sides,

and _____ corners.

They arc _____

Puzzle Corner

Divide this shape, using one line, into a triangle and a pentagon (five-sided shape).

Playing with Shapes

Cut out the shapes. <u>Hint</u>: if you have the download version of this curriculum, print the page of cut-out shapes in landscape orientation, scaled at 140-150%, so it prints on two sheets of paper. All the shapes will then be much bigger.

1. Make a big triangle with the four yellow triangles (marked with 1).

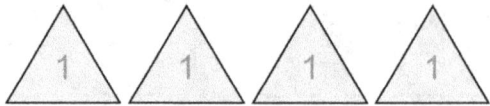

2. Take all six of the yellow triangles (marked with 1). Put them together to get a six-sided shape (a *hexagon*).

3. Use the two pink rectangles (marked with 2) to make a square.

4. Use one pink rectangle (#2) and two blue squares (#7) to make a square.

5. Can you make a bigger square than the one you made in exercise 4? Use any pieces you choose.

6. Make a rectangle using two red triangles (#5).

7. Make a bigger rectangle using four red triangles (#5).

8. Put together two of the green triangles (#4) to get a four-sided shape. You can do this in several different ways!

9. Put together the two slim rectangles (#3) to make **a.** a rectangle; **b.** an L-shape; **c.** an eight-sided shape.

10. Put together the two purple shapes (#6) to make a six-sided shape (a hexagon). You can do this in several different ways!

11. Put together the two purple shapes (#6) to make a four-sided shape (a *quadrilateral*).

12. A challenge: put together the two purple shapes (#6) to make a five-sided shape (a *pentagon*).

13. Make your own pentagons (five-sided figures) using any of the shapes! Make several different ones.

14. Make your own hexagons (six-sided figures) using any of the shapes! Make several different ones.

15. Make interesting figures of your own using any of the shapes. Have fun!

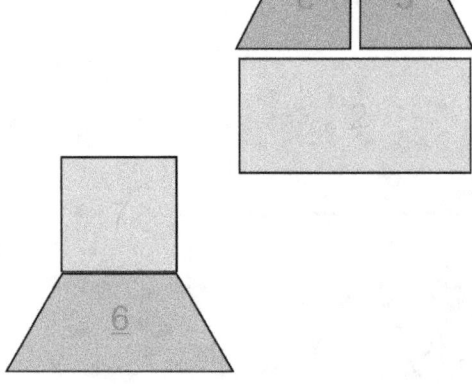

This page intentionally left blank.

Drawing Basic Shapes

1. Use a ruler to join the dots <u>carefully</u> with straight lines. What shape do you get?

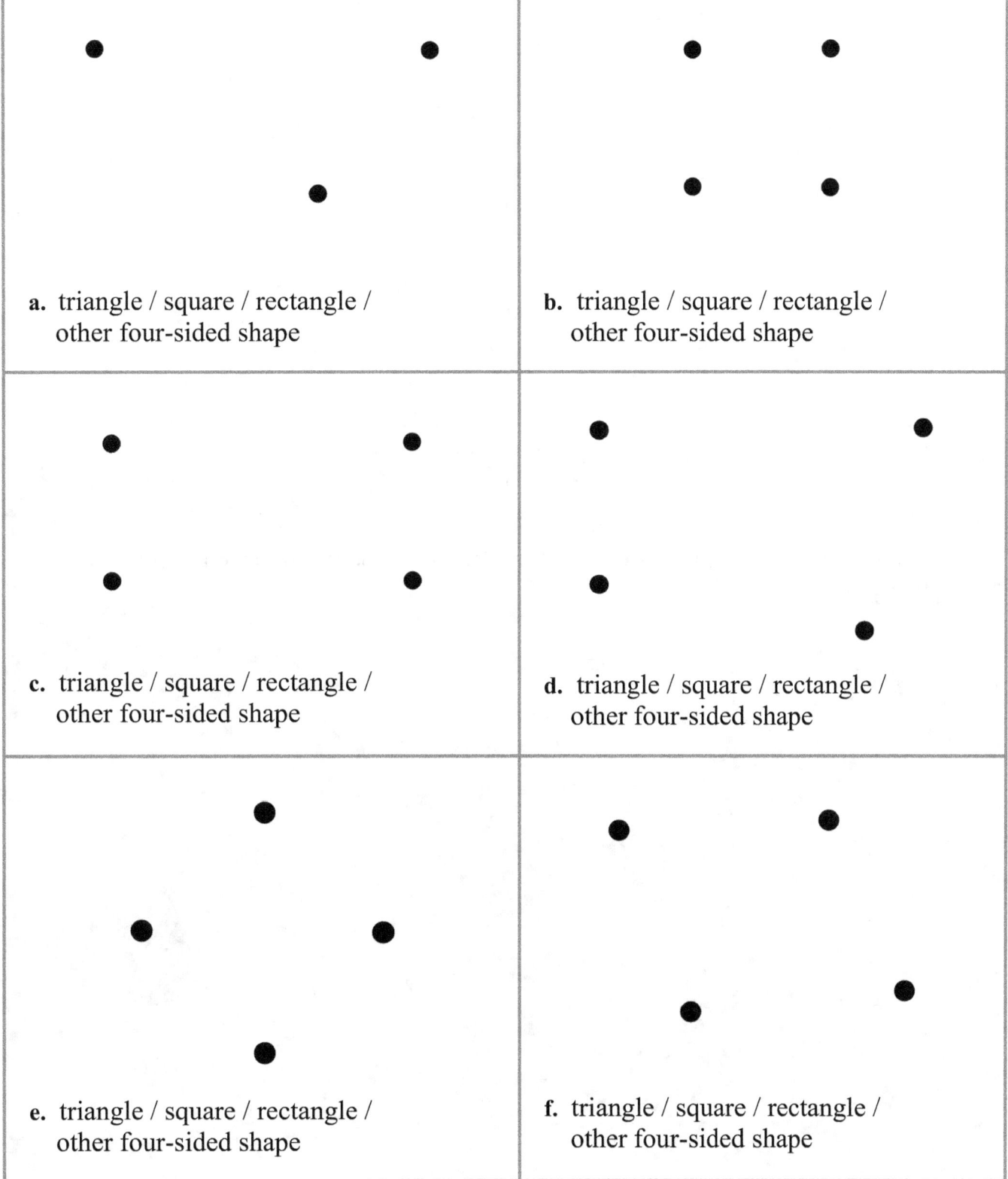

a. triangle / square / rectangle / other four-sided shape

b. triangle / square / rectangle / other four-sided shape

c. triangle / square / rectangle / other four-sided shape

d. triangle / square / rectangle / other four-sided shape

e. triangle / square / rectangle / other four-sided shape

f. triangle / square / rectangle / other four-sided shape

2. Draw.

a. Draw four dots anywhere in this space. Join the dots with lines. Use a ruler!

What shape did you get? A square, a rectangle, or other four-sided shape?

b. In this space try to draw four dots so that you get a rectangle.

c. Draw a rectangle. This time, use something (such as a book) to make the corners perfectly square.

3. Figures (a), (b), (c), and (d) below are all quadrilaterals (four-sided shapes).
   In each shape, draw a line from one corner to the opposite corner.

   What kind of shapes do you get now? _____

   Now draw another line from corner to corner in each shape,
   using the two other corners you didn't yet use.

   How many parts does each four-sided shape have now? _____

   What kind of shapes are these parts? _____

   a.

   b.   **c.**   **d.**

4. Choose a color for each kind of shape, and color them!

   Triangles are _____.   Circles are _____.

   Squares are _____.   Rectangles are _____.

   Other four-sided shapes are _____.

Practicing Basic Shapes and Patterns

1. In each figure, draw a straight line <u>with a ruler</u> from one black dot to the other black dot. Color the two new shapes with different colors. Inside each new shape write a letter: S if it's a square, T if it's a triangle, R if it's a rectangle, Q if it's another quadrilateral (four-sided shape).

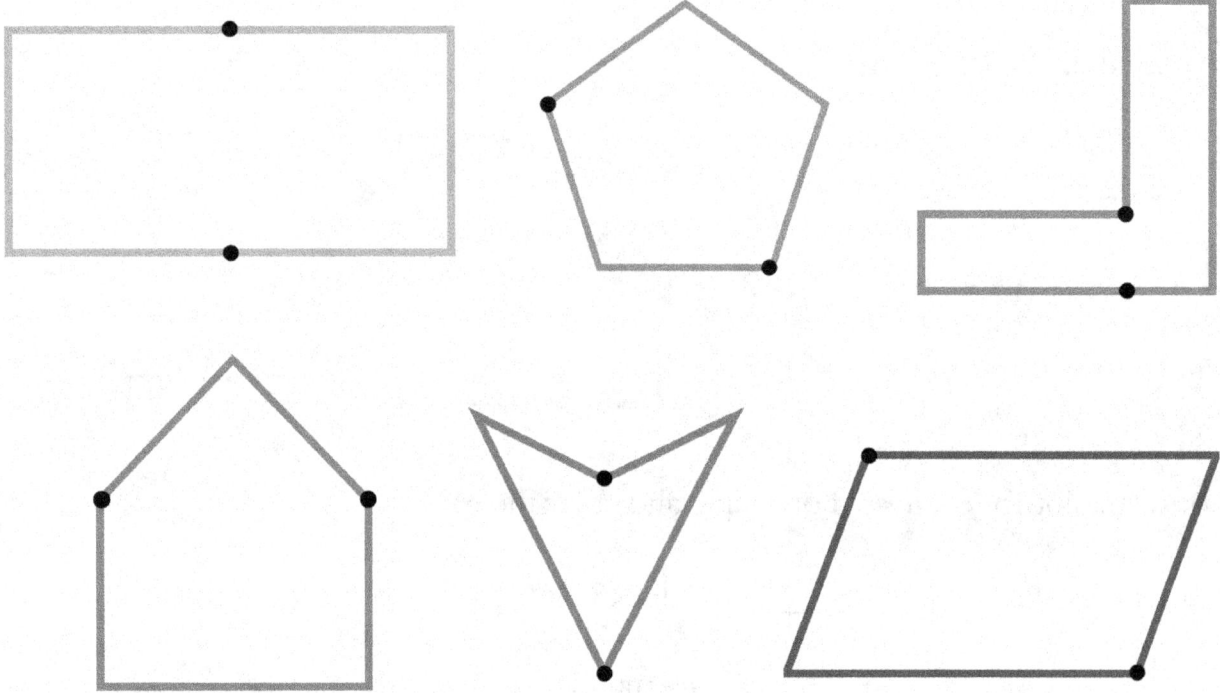

2. Join each dot to a dot on the other side with straight lines (horizontal and vertical lines) so that you get a grid of squares. <u>Use a ruler</u> and draw neatly.

Then color the squares using this pattern (ye = yellow):

blue	green	blue	green
	ye	purple		ye	purple
pink		pink	

3. Repeat the patterns to fill the grids.

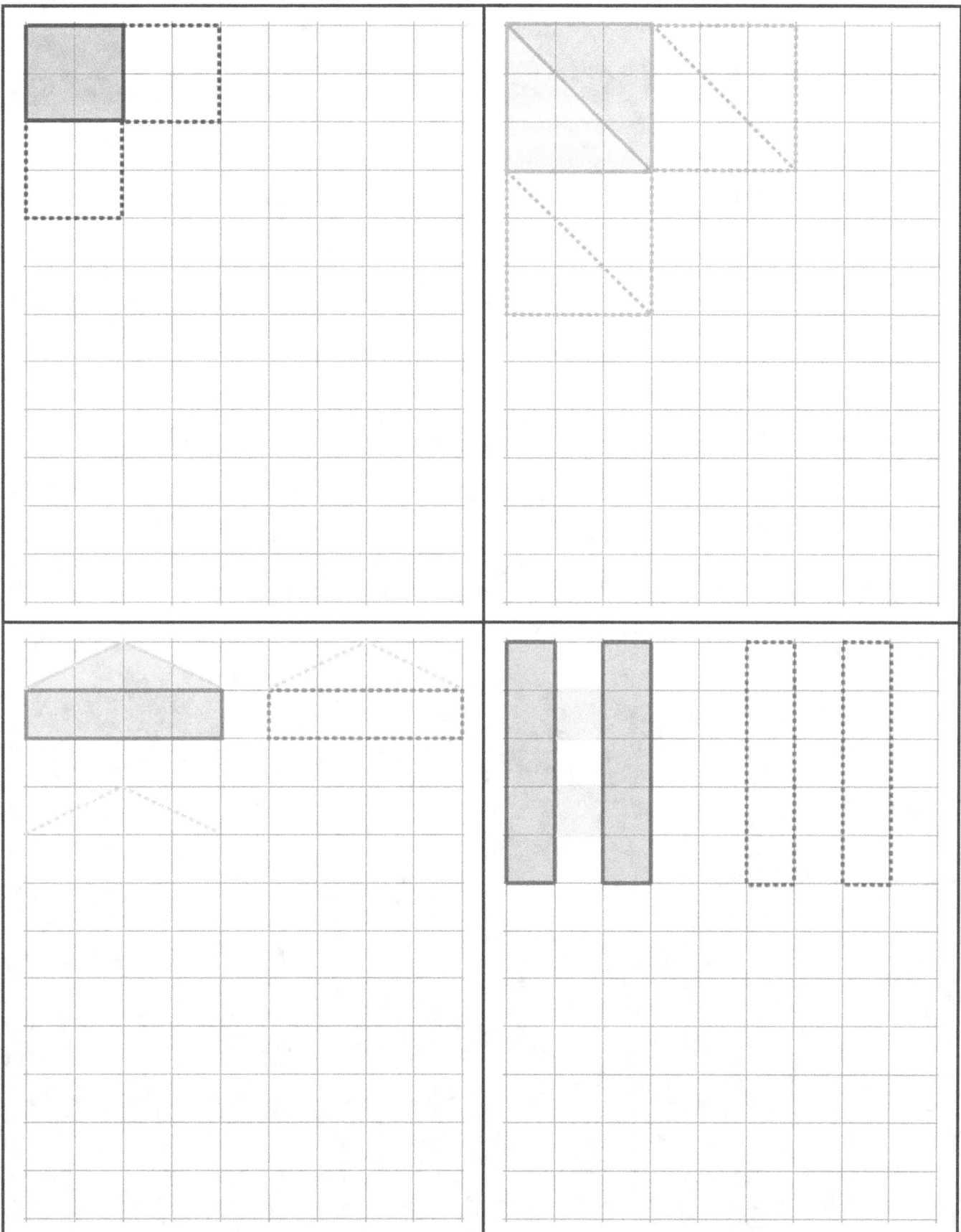

4. Here you can design your own patterns!

Halves and Quarters

This square is divided into two parts that are the same. The parts are **halves**. Each part is one half.

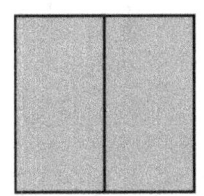

This circle is divided into four parts that are the same. The parts are called **fourths** or **quarters**. Each part is one fourth or one quarter.

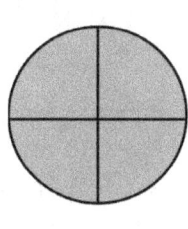

Here, one-half of the square is colored. The other half is white.

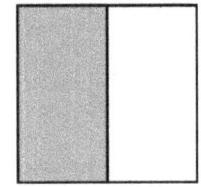

Here, three-fourths of the circle is colored. One-fourth of it is white.

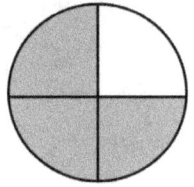

1. Divide these shapes into halves by drawing a straight line from dot to dot. Then color them as the instructions say.

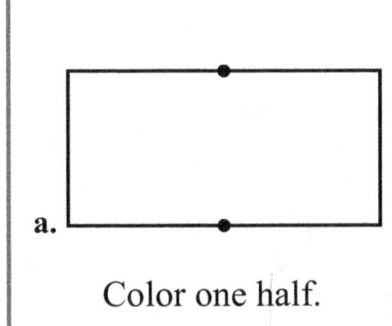

a. Color one half.

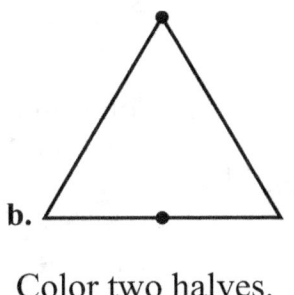

b. Color two halves.

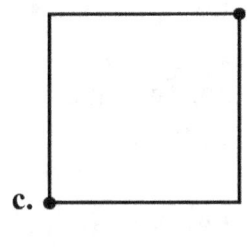

c. Color one half.

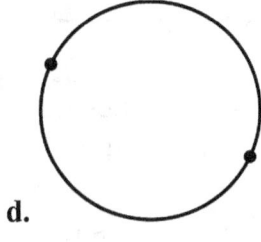

d. Color both halves, but different colors.

2. Divide these shapes into fourths by drawing two straight lines from dot to dot. Then color them as the instructions say.

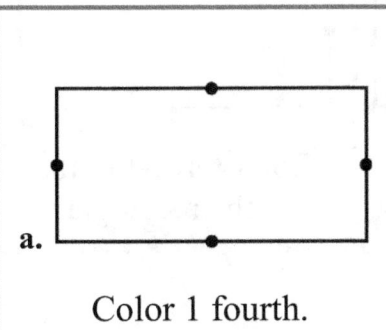

a. Color 1 fourth.

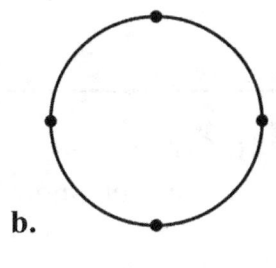

b. Color 3 fourths.

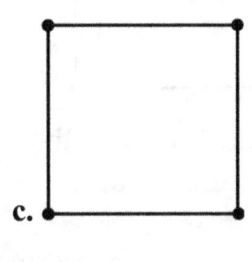

c. Color 2 fourths.

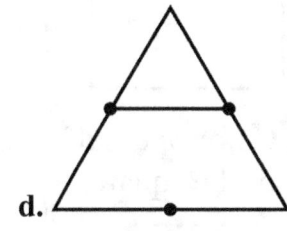

d. Color 4 fourths = the WHOLE triangle.

3. Color. Then compare.

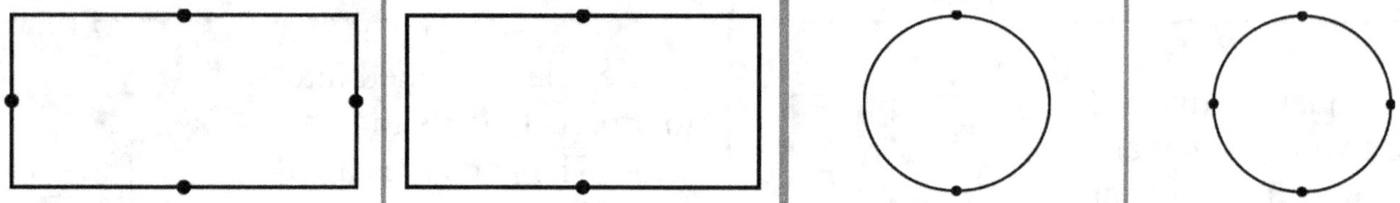

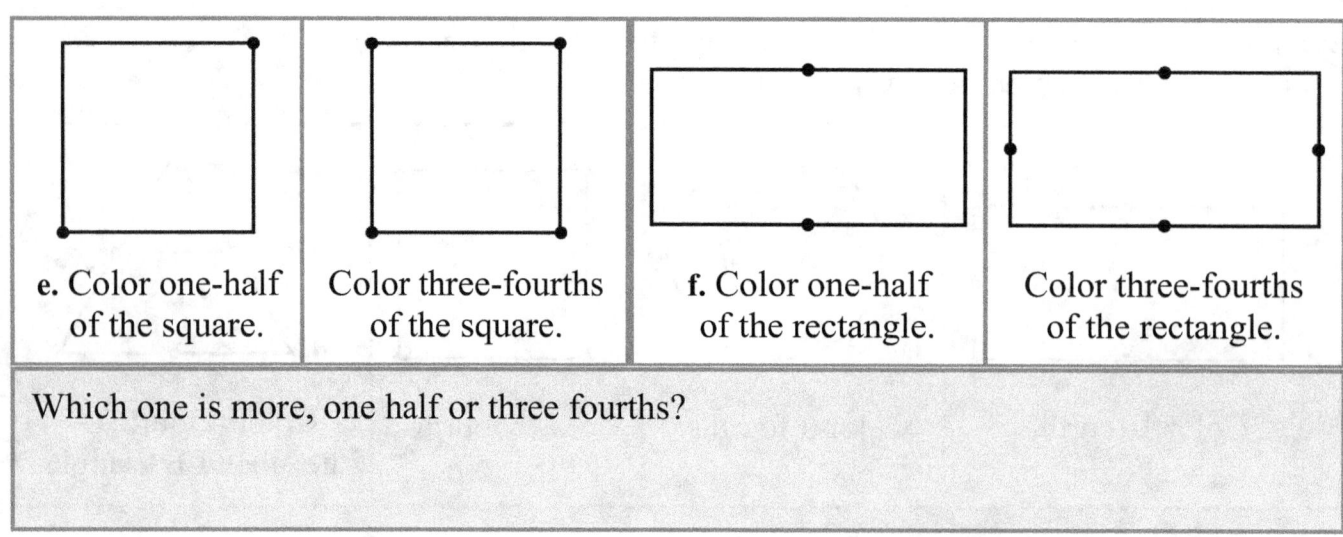

a. Color one-fourth
of the rectangle.

Color one-half of
the rectangle.

b. Color one-half
of the circle.

Color one-quarter
of the circle.

Which one is a bigger piece, one half or one fourth?

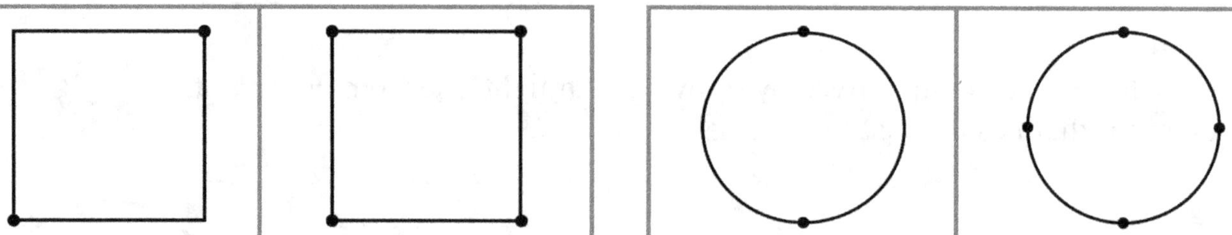

c. Color one-half
of the square.

Color two quarters
of the square.

d. Color one-half
of the circle.

Color two quarters
of the circle.

Which one is more, one half or two quarters?

e. Color one-half
of the square.

Color three-fourths
of the square.

f. Color one-half
of the rectangle.

Color three-fourths
of the rectangle.

Which one is more, one half or three fourths?

This square is divided into three parts that are the same.
The parts are **thirds**. Each part is <u>one third</u>.

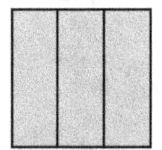

4. Color.

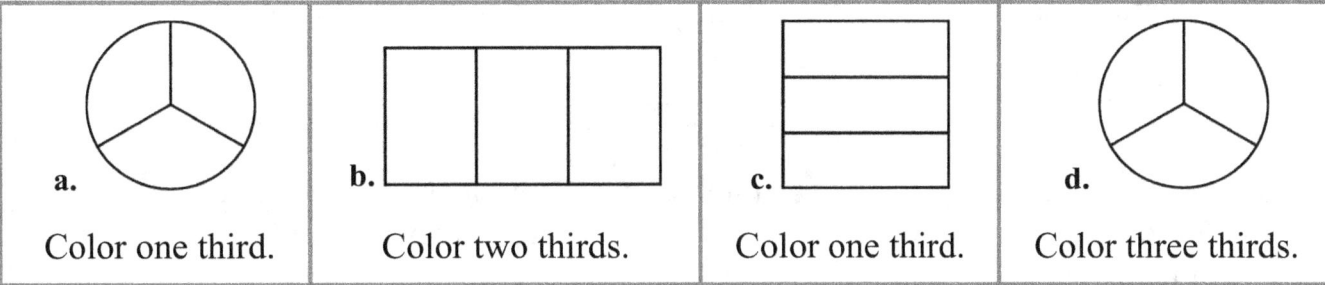

a. Color one third.	**b.** Color two thirds.	**c.** Color one third.	**d.** Color three thirds.

5. Color. Then compare.

 Color two thirds.	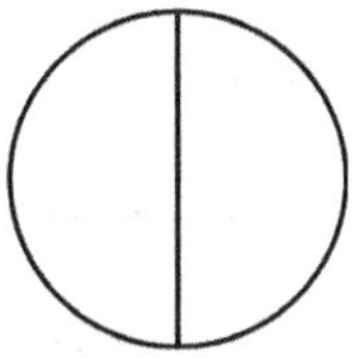 Color one half.

a. Which is more, two thirds or one half?

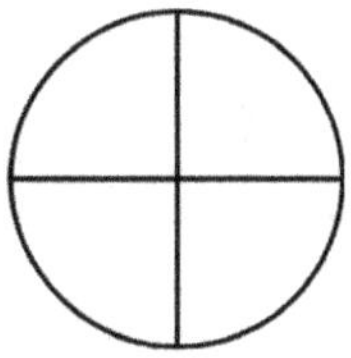 Color three fourths.	 Color two thirds.

b. Which is more, three fourths or two thirds?

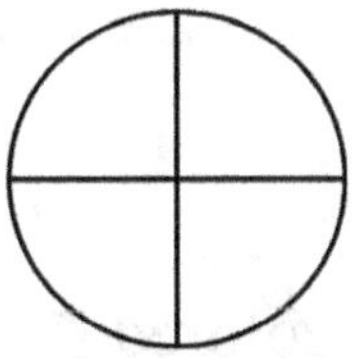

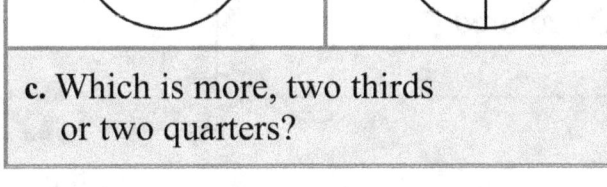

c. Which is more, two thirds or two quarters?

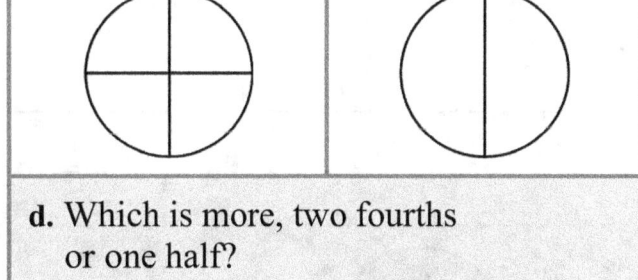

d. Which is more, two fourths or one half?

6. Color ONE piece in each pie. Then compare. Think of eating pie pieces!

a. Which is more, one half or one third?

b. Which is more, one fourth or one third?

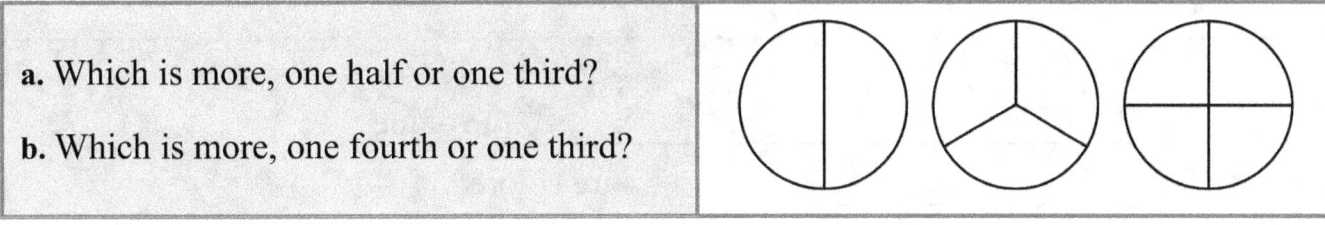

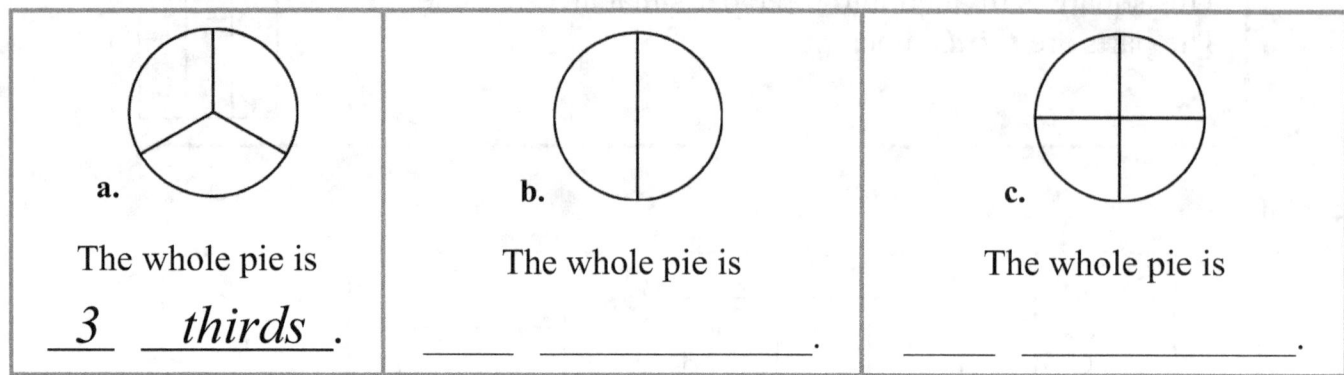

7. Color the whole pie. Then tell or write how many pieces it is, and what kind of pieces.

a.	b.	c.
The whole pie is	The whole pie is	The whole pie is
__3__ __thirds__ .	_____ _____ .	_____ _____ .

8. Complete these sentences like the example so that they say how many pieces
   are colored, what kind of pieces they are, and what shape they belong to.
   Look at the example.

a. ___1___ _____fourth_____ of

the _____oval_____ is colored.

b. _____ _____ of

the _____hexagon_____ is colored.

c. _____ _____ of the

_____trapezoid_____ is colored.

d. _____ _____ of the

_____ are colored.

e. _____ _____ of the

_____ are colored.

f. _____ _____ of the

_____ are colored.

72

Measuring Length

In this lesson, we measure things to find <u>how long</u> or <u>how wide</u> they are as compared to other things. For all measuring, you need a **measuring unit**. You <u>repeat</u> the measuring unit many times, and compare it to the thing you are measuring.

1. Measure <u>how wide</u> or <u>how long</u> things are, using shoes as measuring units.
   <u>**You need**</u>: two small shoes and two bigger shoes.

   a. Measure a desk or a table. Place one shoe at the edge
   of the table and the other one directly behind it. Then
   move the first shoe in front of the second, and so on.
   Keep count.

   The table is _____ small shoes wide.

   The table is _____ big shoes wide.

   b. Measure two more things now, using both the small shoes and the big shoes.
   Some ideas:

   • how wide the blackboard is;   • how long your friend is when lying on the floor;
   • how tall your chair is;   • how long the room is.

   The _____ is _____ small shoes wide.

   The _____ is _____ big shoes wide.

   The _____ is _____ small shoes wide.

   The _____ is _____ big shoes wide.

2. Ryan noticed that each daddy shoe was about <u>three</u> baby shoes.
   Ryan measured his desk and it was four daddy-shoes wide, like this:

   How many baby-shoes wide is Ryan's desk?
   Hint: Draw the baby shoes under the daddy shoes.

3. Ryan measured that his room was 27 shoes wide, using daddy shoes.
He also measured it using baby shoes.
Was Ryan's room 81 baby-shoes wide, or 9 baby-shoes wide?

4. Measure how long some things are, using paperclips.
You need: several paperclips that are the same size,
small things to measure such as an eraser, a pencil,
crayons, toys, or books.

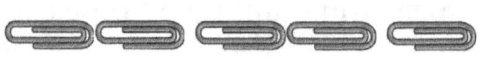

Write the things below **in order**, from shortest to longest.

_____   _____ paper clips

_____   _____ paper clips

_____   _____ paper clips

_____   _____ paper clips

_____   _____ paper clips

5. How many crayons long are these pencils? How many paper clips long are they?

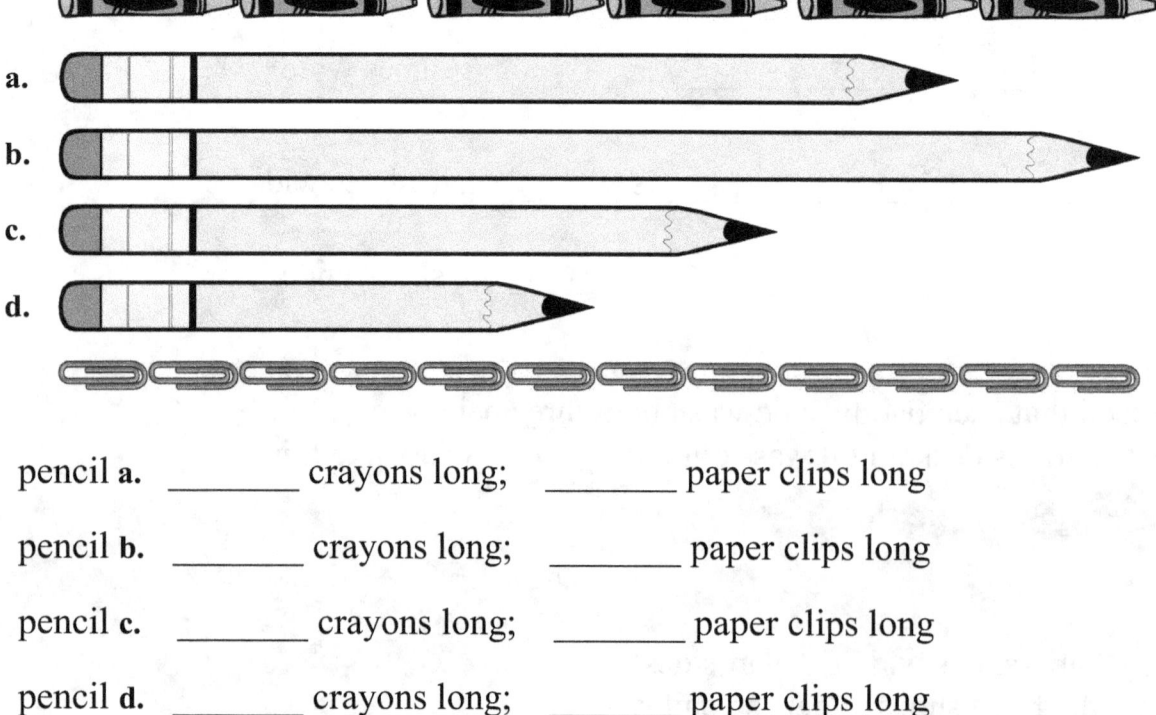

pencil **a.** _____ crayons long; _____ paper clips long

pencil **b.** _____ crayons long; _____ paper clips long

pencil **c.** _____ crayons long; _____ paper clips long

pencil **d.** _____ crayons long; _____ paper clips long

Sometimes we cannot easily tell which of two things is longer or wider. We can use a third thing as a "measuring stick." Look at these two houses. Can you tell which one is longer?

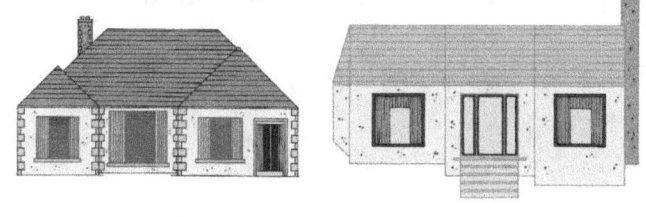

Now let's use this "log" as a measuring stick:

   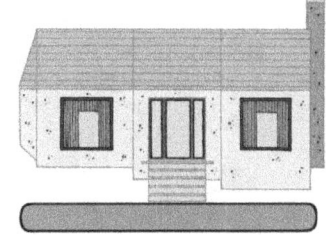

House 1 is a little shorter than our log.   House 2 is a little longer than our log.

Is house 1 longer than house 2? Or the other way around?

House 2 is longer than house 1, because it is longer than our log, whereas house 1 is shorter than the log.

6. Compare the things to the "measuring stick." Mark the longer of the two.

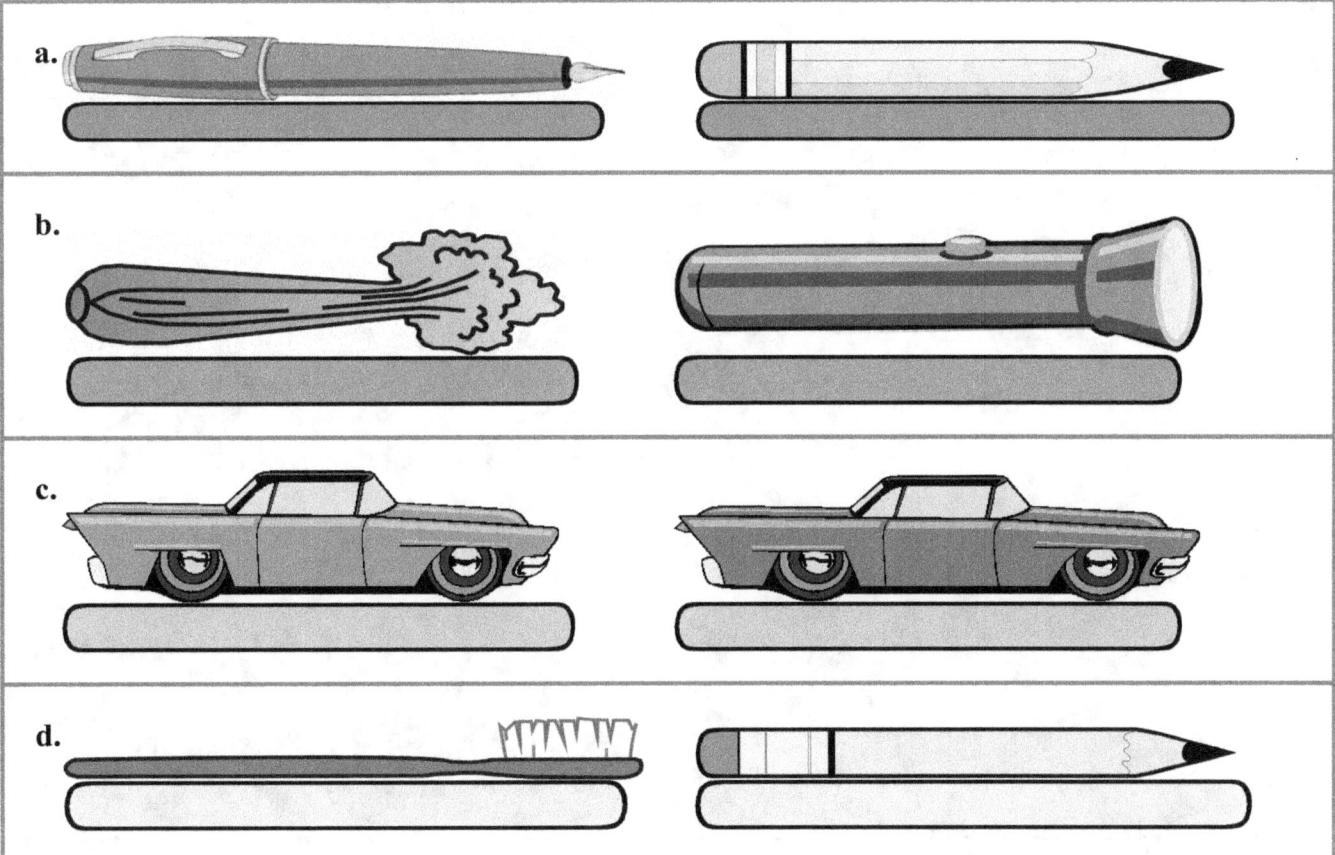

7. Draw a picture to match the situation. You can draw stick figures.

a. Jerry is shorter than the top of the cabinet. The top of the cabinet is shorter than Mike.

b. The table is taller than little Kyle. Little Mary is taller than the table.

Exploring Measuring

Besides measuring length, we also measure things to find how heavy something is, how much liquid it holds, or how much space it takes, as compared to other things.

1. Find <u>five</u> things you can carry, some lighter and some heavier. Put them in order from the lightest to the heaviest. You can draw the things or write them in the space below.

2. Order these things from lightest to heaviest by writing 1, 2, and 3 next to them. Don't just go by which picture looks bigger. Think how heavy these things would be in real life.

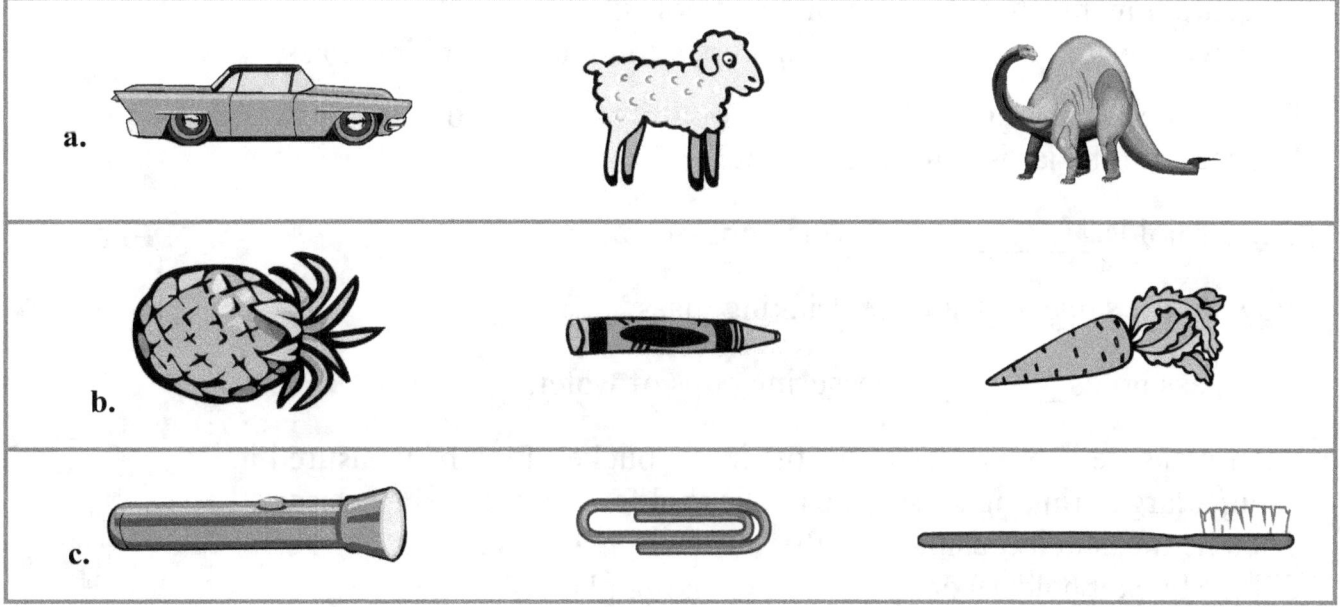

a.

b.

c.

3. If you have a bathroom scale, step on it and measure how much you weigh. Weigh some other things, also. If your scale measures in pounds, write "lb" after the number, such as 45 lb. If it measures in kilograms, write "kg", such as 22 kg.

I weigh _____ _____.

_____ weighs _____ _____.

_____ weighs _____ _____.

_____ weighs _____ _____.

> For all measuring, you need a **measuring unit**. You <u>repeat</u> the measuring unit a lot of times to compare it to the thing you are measuring.

4. Measure how much water a pot holds.
   You need: water, a large coffee cup, a food jar, and a pot or other big container.

   Fill the cup with water and pour into the pot. Repeat until the pot is full. Keep count of how many cups full of water you need to fill the pot.

   The pot holds _____ cups full of water.

   Now do the same using a jar:

   The pot holds _____ jars of water.

5. Measure how much water a jar or a cup holds.
   You need: water, a small measuring cup, a food jar, a drinking glass.

   Fill the measuring cup with water and pour it into the food jar.
   Repeat until the jar is full. Keep count.

   The jar holds _____ measuring cups of water.

   Now do the same with a large drinking glass.

   The glass holds _____ measuring cups of water.

6. Peter measured how much water fits into a bucket. First he measured it using a large drinking glass. The bucket holds 32 big drinking glasses. Then he measured it using a <u>smaller</u> drinking glass. Which is correct: did the bucket hold 19 or 53 smaller drinking glasses?

Measuring Lines in Inches

This line is 1 inch long. ├────────┤

We also write "1 in." for short.

1. How many inches are end-to-end?

a. ├────┼────┤ _____ inches

b. ├────┼────┼────┼────┤ _____ inches

c. ├────┼────┼────┤ _____ inches

d. ├────┼────┼────┼────┤ _____ inches

2. How many inches long are these items?

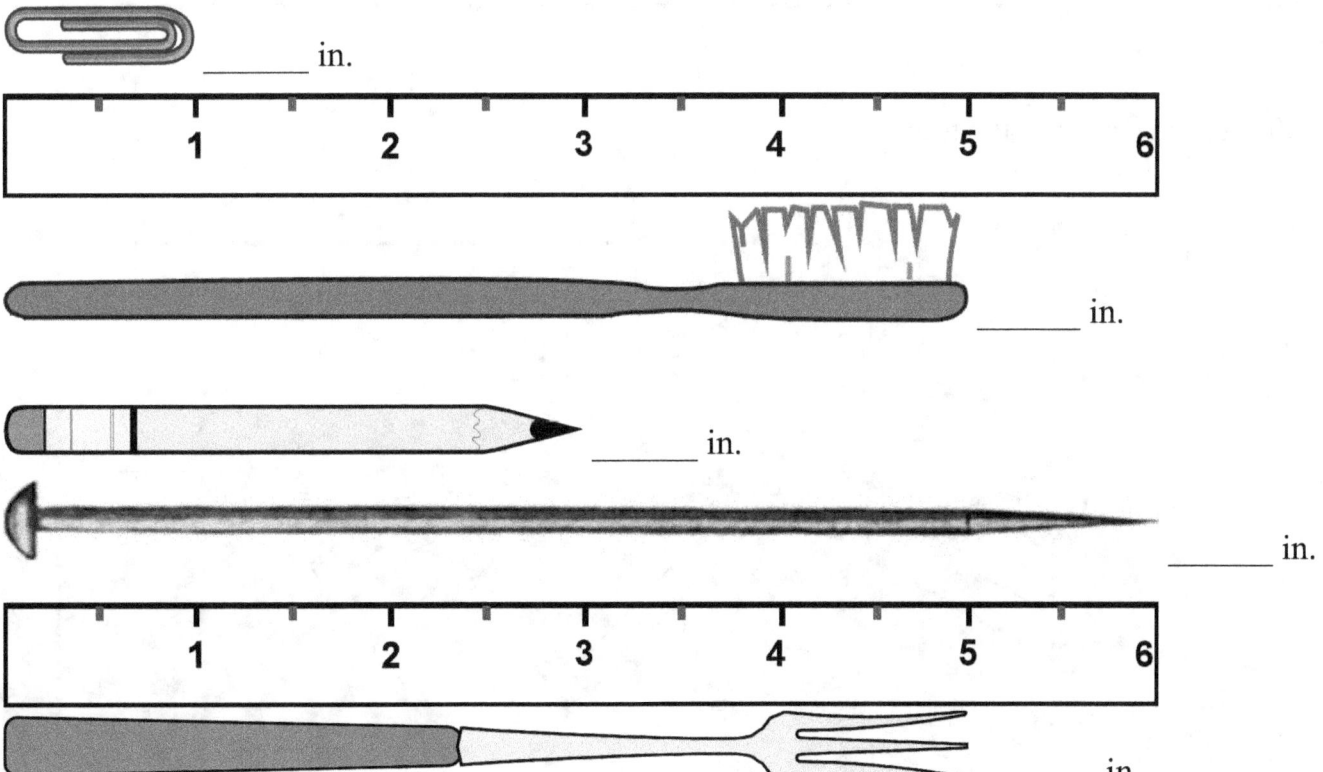

_____ in.

_____ in.

_____ in.

_____ in.

_____ in.

3. Measure the lines with a ruler.

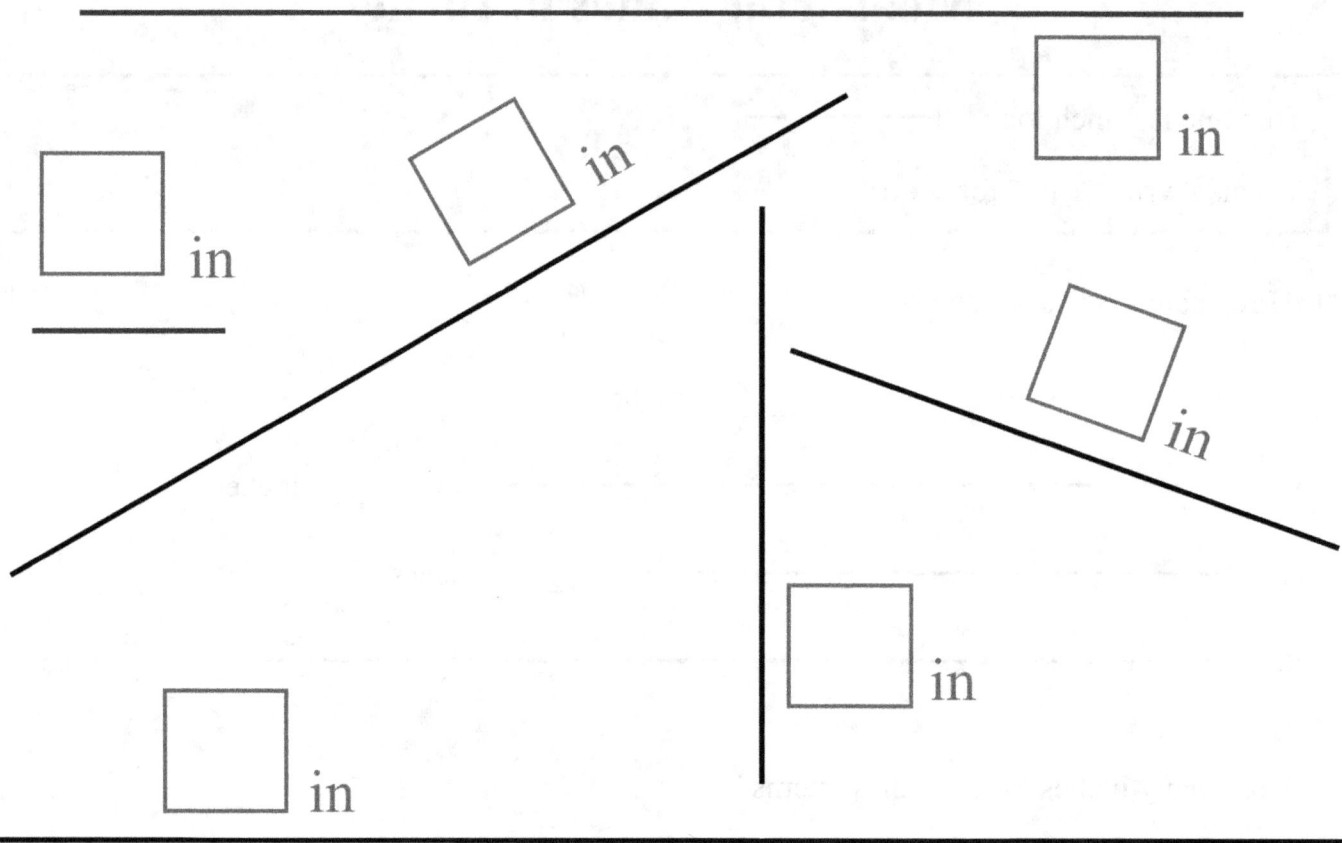

4. Measure the sides of the triangles.

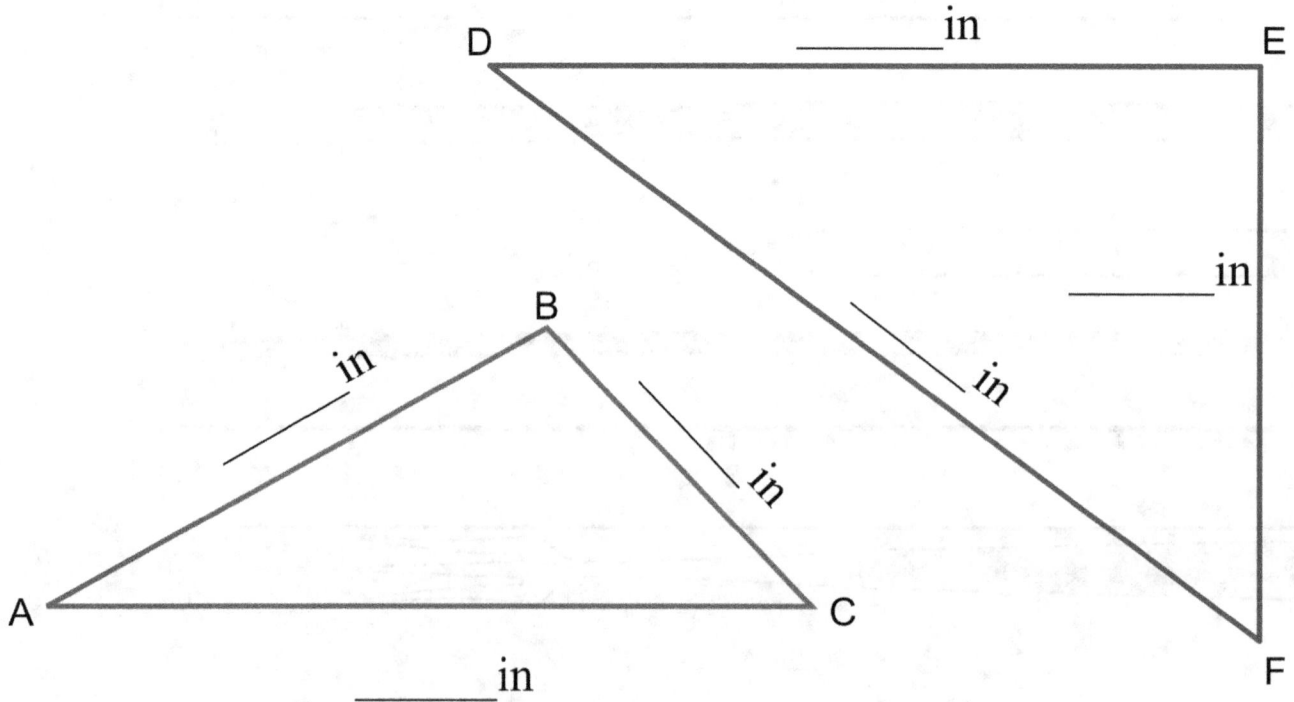

5. Use a ruler and draw lines with these lengths:

   a. 4 in.

   b. 2 in.

   c. 5 in.

   d. 7 in.

   e. 1 in.

   f. 8 in.

6. Draw the last side for these figures with a ruler. Then measure all the sides of each figure. Write the measurement next to each side (for example "2 inches" or "2 in.").

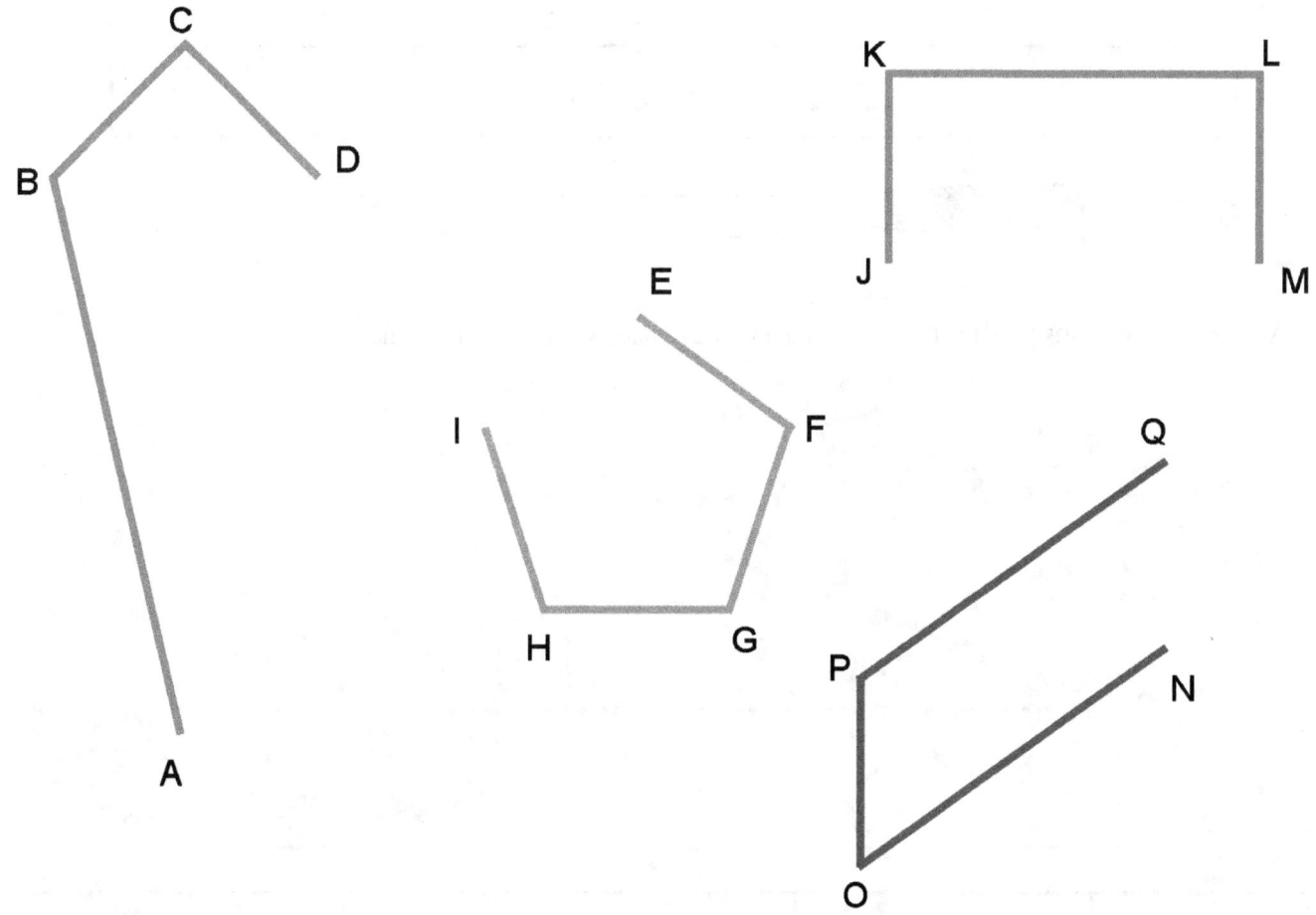

Measuring Lines in Centimeters

You can find out how long things are in *centimeters*.

This line is 1 centimeter long: ⊢──┤
A centimeter is written in its short form as "cm."
This pencil is 6 cm long.

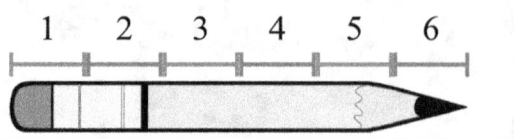

1. How many centimeters long are these things?

a. _____ cm

b. _____ cm

c. _____ cm

d. _____ cm

e. _____ cm

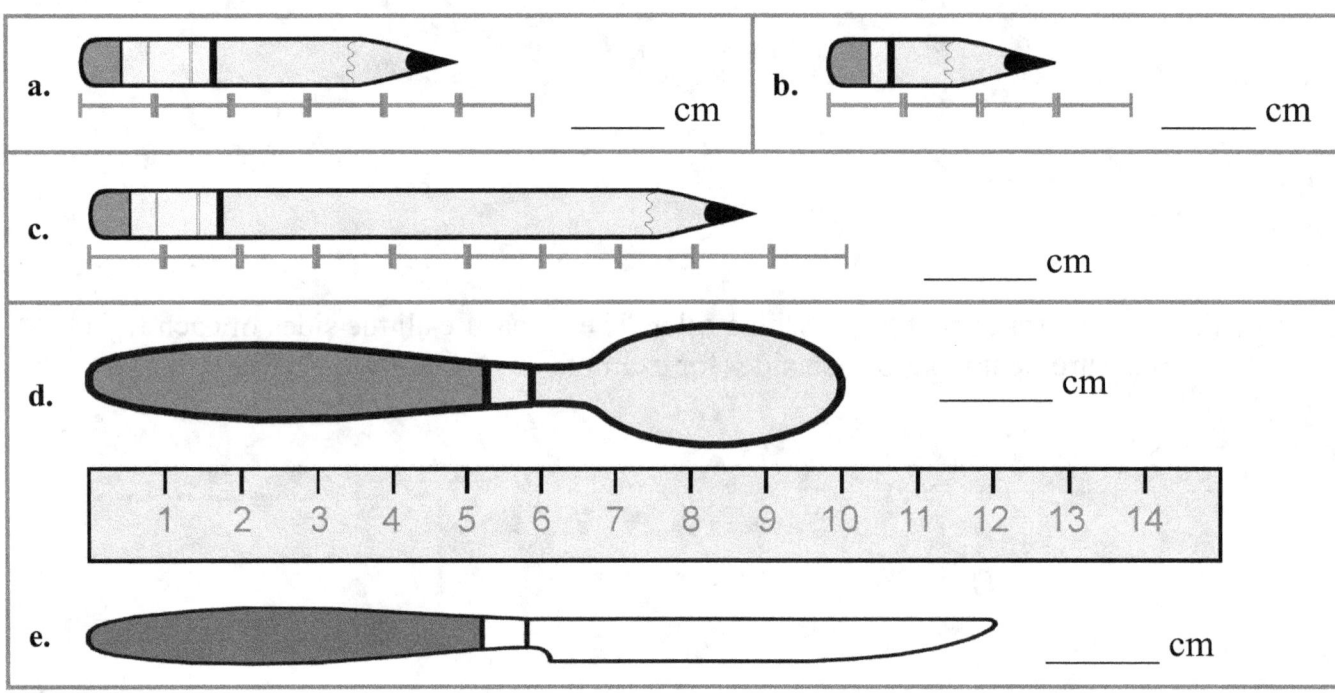

2. Measure the lines with a ruler. (If you don't have one, cut out the ruler at the bottom of the page.)

☐ cm

☐ cm

☐ cm

☐ cm

☐ cm

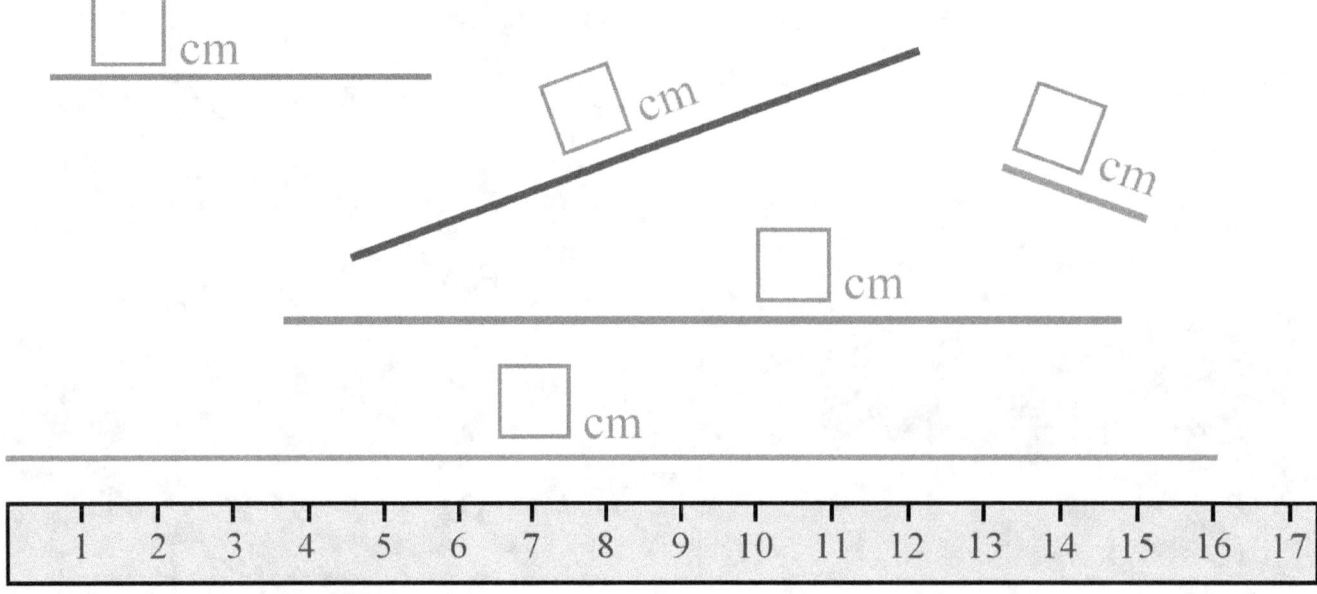

3. Draw the last side for these figures with a ruler. Then measure all three sides of each figure. Write the measurement next to each line (for example "6 cm").

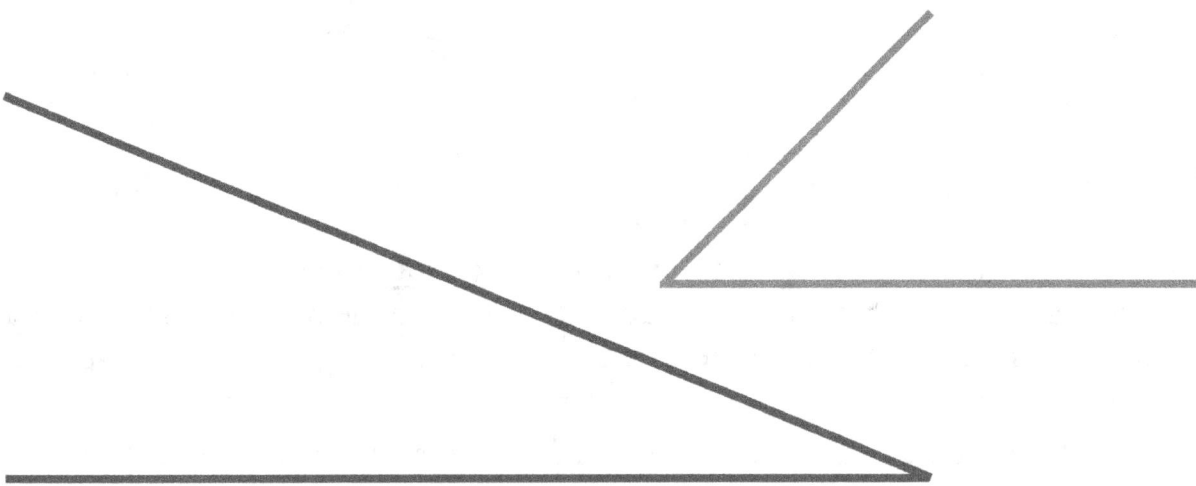

4. Use your own ruler and draw lines that are these lengths.

   a. 4 cm

   b. 5 cm

   c. 8 cm

   d. 16 cm

5. Measure some things around you! For example, a book, your pencil, a table, and so on.

Thing	How long?

Three-Dimensional Shapes

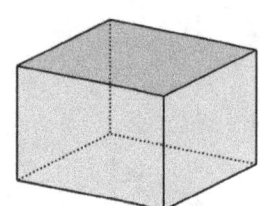 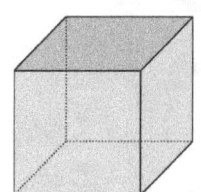 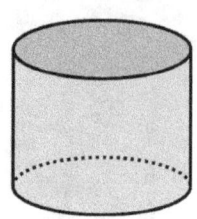 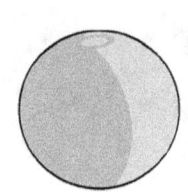

This is a **box**. It is also called a "rectangular prism."

A **cube** is a box, too, but all of its edges are the same length.

A **cylinder** has a circle on the bottom and on the top.

This is a **ball**, or a sphere.

1. Are these things in the shape of a *box* or a *cube*? Underline the right choice.

a.

box *or* cube

b.

box *or* cube

c.

box *or* cube

d.

box *or* cube

e.

box *or* cube

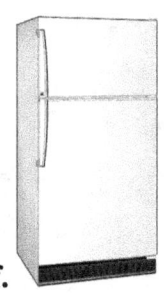

f.

box *or* cube

g.

box *or* cube

h.

box *or* cube

2. Find four things in your classroom or at home in the shape of a *box*.
   Put them in order from the smallest to the biggest.

   I found _____, _____,

   _____, and _____.

3. Find two things in your classroom or at home in the shape of a *cube*,
   one smaller and one bigger.

   I found _____ and _____.

4. Are these things in the shape of a *cylinder* or a *ball*? Underline the right choice.

   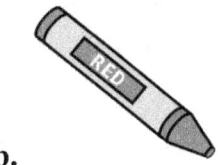   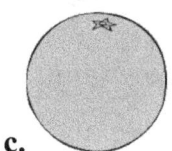   

a.   **b.**   **c.**   **d.**

cylinder *or* ball   cylinder *or* ball   cylinder *or* ball   cylinder *or* ball

             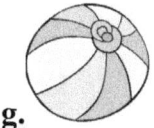   

e.   **f.**   **g.**   **h.**

cylinder *or* ball   cylinder *or* ball   cylinder *or* ball   cylinder *or* ball

5. Which shapes can roll on the floor? Underline. *cylinder   box   ball   cube*

6. Which shapes will slide on the floor and not roll?   *cylinder   box   ball   cube*

7. Find four things in your classroom or at home in the shape of a *ball*.
   Put them in order from the smallest to the biggest.

   I found _____, _____,

   _____, and _____.

8. Find four things in your classroom or at home in the shape of a *cylinder*.
   Put them in order from the smallest to the biggest.

   I found _____, _____,

   _____, and _____.

9. Name the basic shape.

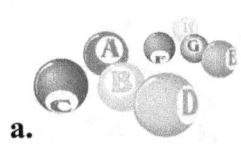      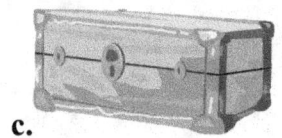   

a.   **c.**   **d.**

b.

Review Chapter 6

1. Divide the shapes using one straight line.

   Divide Shape A into a triangle and a five-sided shape.

   Divide Shape B into a square and a rectangle.

   Divide Shape C into a four-sided shape and a triangle.

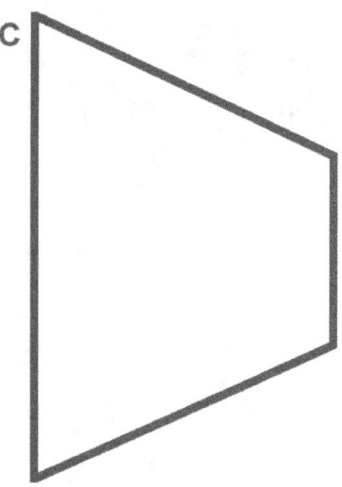

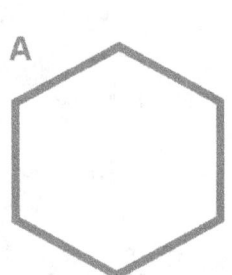

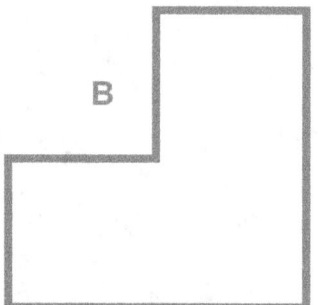

2. Color the triangles orange,
   the rectangles red,
   the squares blue, and
   the little circles light blue.

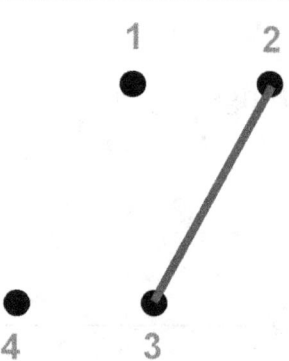

3. Join these dots <u>carefully</u> with lines, from 1 to 2 to 3 to 4 to 1. Use a ruler.

   What shape do you get?

   Divide your shape into two triangles.

4. How many corners does this shape have?

   (We call it a *pentagon*.)

   Measure its sides in centimeters.

Chapter 7: Adding and Subtracting Within 0-100
Introduction

The seventh chapter progressively presents a variety of easy addition and subtraction problems with numbers from 0 to 100. It includes these topics:

- Adding a two-digit number and a single-digit number without regrouping (carrying) (for example, 23 + 4 or 56 + 3).

- Subtracting a one-digit number from a two-digit number without regrouping (borrowing): For example, 45 − 3 or 67 − 6.

- Adding or subtracting two-digit numbers in columns (one number under the other) without regrouping.

- Recognizing that sometimes in adding two-digit numbers we need to regroup — to combine ten ones to make a new ten. In this grade level, we approach this concept using visual models only, and not in an abstract manner.

- Learning specific mental math strategies for adding and subtracting numbers under 20 (such as 7 + 9 or 15 − 8). We study a trick with nine and eight, adding just one more than a known sum, and using the relationship between addition and subtraction to subtract. The actual memorization of the basic addition and subtraction facts within 0-18 is left for second grade.

Pacing Suggestion for Chapter 7

Please add one day to the pacing for the test if you will use it. Note that the specific lessons in the chapter can take several days to finish. They are not "daily lessons." As a general guideline, first graders should finish 1-2 pages daily or 7-9 pages a week. Please also see the user guide at https://www.mathmammoth.com/userguides/ .

The Lessons in Chapter 7	page	span	suggested pacing	your pacing
Refresh Your Memory	90	*2 pages*	1 day	
Adding Without Carrying	92	*3 pages*	2 days	
Subtracting Without Borrowing	95	*3 pages*	2 days	
Adding or Subtracting Two-Digit Numbers	98	*4 pages*	2 days	
Completing the Next Ten	102	*3 pages*	2 days	
Going Over Ten	105	*4 pages*	2 days	
Subtracting from Whole Tens	109	*2 pages*	2 days	
Add Using "Just One More".............................	111	*2 pages*	2 days	
A "Trick" with Nine and Eight	113	*3 pages*	2 days	
Adding Within 20	116	*4 pages*	2 days	
Subtract to 10	120	*2 pages*	1 day	
Using Addition to Subtract	122	*3 pages*	2 days	
Some Mixed Review	125	*3 pages*	2 days	
Pictographs	128	*2 pages*	1 day	
Review Chapter 7	130	*4 pages*	2 days	
Chapter 7 Test (optional)				
TOTALS		*44 pages*	27 days	

Games and Activities at Math Mammoth Practice Zone

Hidden Picture Addition Game
Use a number range of 3 to 19, or some other, to practice addition.
https://www.mathmammoth.com/practice/mystery-picture

Hidden Picture Subtraction Game
Choose a number range of 2 to 18, for example, to practice subtraction in this fun game.
https://www.mathmammoth.com/practice/mystery-picture-subtraction

Bingo
Choose Addition (Single-Digit).
https://www.mathmammoth.com/practice/bingo

Mathy's Berry Picking Adventure
Join Mathy (our mammoth mascot) on his berry-picking adventure, and practice your mental math! The link below gives you addition and subtraction problems within 0-20.
https://www.mathmammoth.com/practice/mathy-berries#mode=addition-single&duration=2m

Make Addition Sentences
You're given numbers (in flowers), and an answer to an addition. Drag two flowers to the empty slots so that the addition is true.
https://www.mathmammoth.com/practice/number-sentences#questions=10&types=add-1-20

Two-Digit Mental Addition - Online Practice
Practice adding one two-digit number and one single-digit number without regrouping in this online quiz.
https://www.mathmammoth.com/practice/addition-subtraction-two-digit#opts=2p1dnr

Two-Digit Mental Subtraction - Online Practice
Practice subtracting a single-digit number from a two-digit number without regrouping in this online quiz.
https://www.mathmammoth.com/practice/addition-subtraction-two-digit#opts=2m1dnr

Fruity Math
Click the fruit with the correct answer and try to get as many points as you can in the allotted time.
You could start with Level 1, in which case the game will automatically advance towards harder problems, or you could use the "Manual" setting to choose the exact types of numbers to add.
https://www.mathmammoth.com/practice/fruity-math

Further Resources on the Internet

We have compiled a list of Internet resources that match the topics in this chapter, including pages that offer:

- **online practice** for concepts;
- online **games**, or occasionally, printable games;
- **animations** and interactive **illustrations** of math concepts;
- **articles** that teach a math concept.

We heartily recommend you take a look! Many of our customers love using these resources to supplement the bookwork. You can use these resources as you see fit for extra practice, to illustrate a concept better and even just for some fun. Enjoy!

https://l.mathmammoth.com/gr1ch7

Scan me

Refresh Your Memory

1. Divide (break up) the numbers into two parts.

7	8	9	10
5 and _____	1 and _____	4 and _____	3 and _____
6 and _____	4 and _____	8 and _____	9 and _____
1 and _____	5 and _____	2 and _____	6 and _____
2 and _____	7 and _____	1 and _____	5 and _____
4 and _____	2 and _____	3 and _____	2 and _____

2. Subtract. You can use the number pairs above.

a.	b.	c.	d.
$7 - 1 =$ _____	$8 - 3 =$ _____	$9 - 2 =$ _____	$10 - 3 =$ _____
$7 - 5 =$ _____	$8 - 6 =$ _____	$9 - 3 =$ _____	$10 - 8 =$ _____
$7 - 3 =$ _____	$8 - 2 =$ _____	$9 - 5 =$ _____	$10 - 5 =$ _____
$7 - 6 =$ _____	$8 - 7 =$ _____	$9 - 7 =$ _____	$10 - 6 =$ _____

3. Write a fact family for each picture.

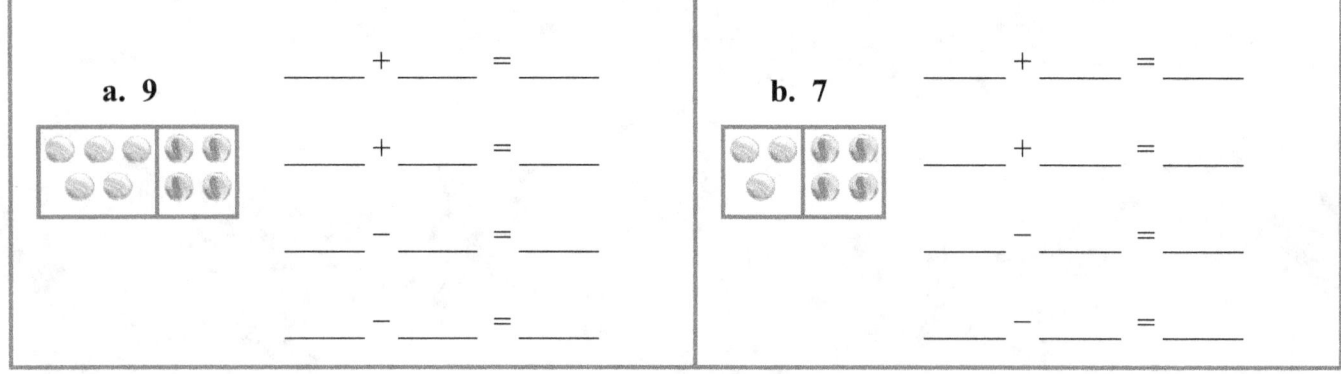

a. 9
_____ + _____ = _____
_____ + _____ = _____
_____ − _____ = _____
_____ − _____ = _____

b. 7
_____ + _____ = _____
_____ + _____ = _____
_____ − _____ = _____
_____ − _____ = _____

4. Add or subtract.

a. $7 + 1 + 2 =$ _____	b. $3 + 5 + 1 =$ _____	c. $10 - 2 - 2 =$ _____
$4 + 1 + 4 =$ _____	$2 + 2 + 4 =$ _____	$7 - 1 - 5 =$ _____
$3 + 1 + 2 =$ _____	$6 - 1 - 2 =$ _____	$9 - 3 - 4 =$ _____

5. Add and subtract. Start with the number in the bottom left corner and follow the arrows.

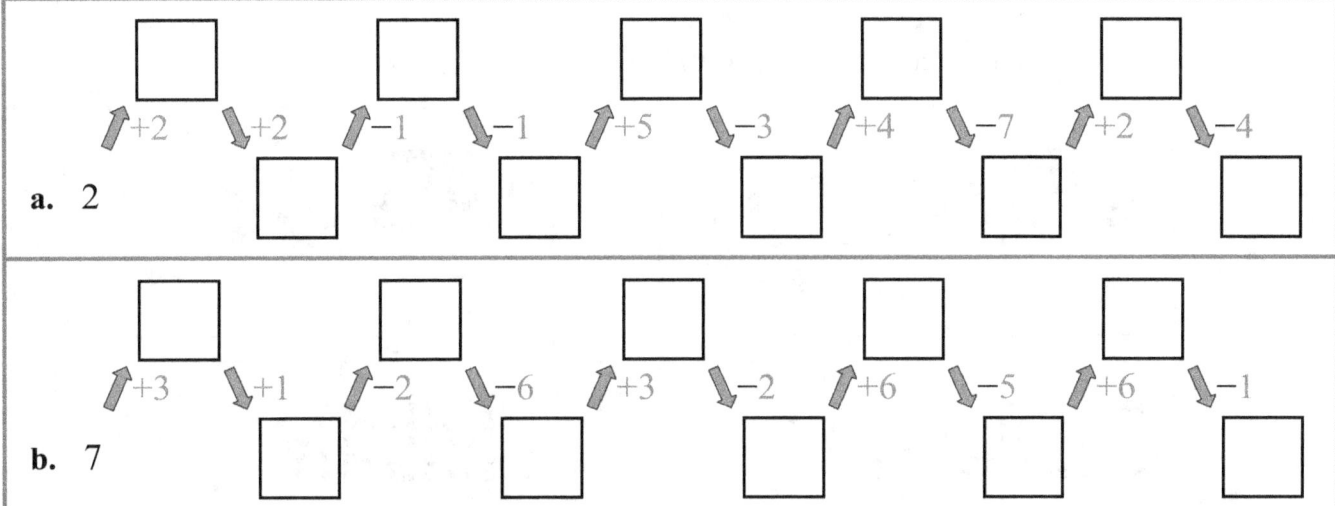

6. Count by tens. If you have forgotten how to do it, a 100-chart can help.

a. 10, 20, _____, _____, _____, _____, _____, _____, _____

b. 24, 34, _____, _____, _____, _____, _____, _____, _____

c. _____, _____, _____, 48, 58, _____, _____, _____, _____

7. Add or subtract.

a.	b.	c.	d.
$20 + 6 =$ _____	$20 +$ _____ $= 24$	$10 + 10 =$ _____	$50 - 10 =$ _____
$40 + 8 =$ _____	$70 +$ _____ $= 78$	$32 + 10 =$ _____	$66 - 10 =$ _____
$50 + 2 =$ _____	$50 +$ _____ $= 56$	$67 + 10 =$ _____	$82 - 10 =$ _____

Adding Without Regrouping

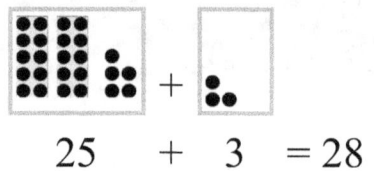
25 + 3 = 28

Add 5 + 3 first.
The 2 tens do not change.

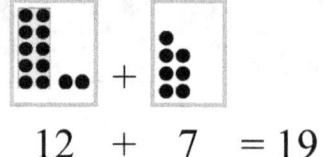
12 + 7 = 19

Add 2 + 7 first.
The ten does not change.

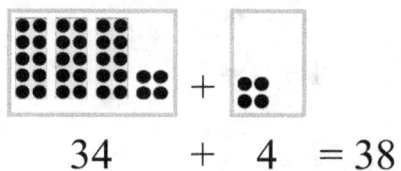
34 + 4 = 38

Add 4 + 4 first.
The 3 tens do not change.

1. Write an addition sentence for each picture.

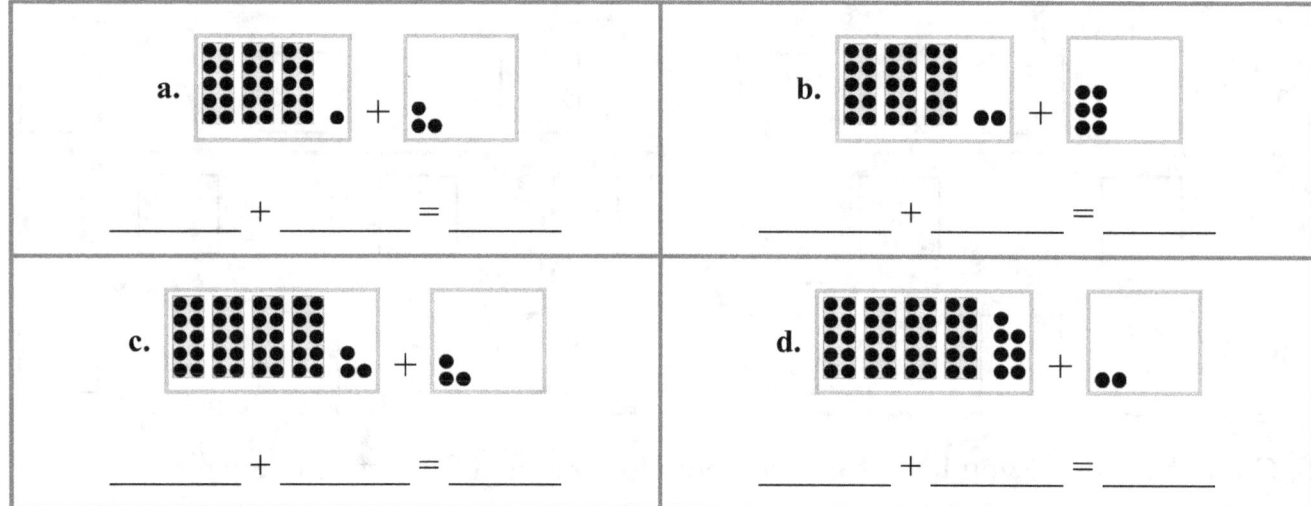

a. _____ + _____ = _____

b. _____ + _____ = _____

c. _____ + _____ = _____

d. _____ + _____ = _____

2. Add. Compare the problems. The top problem helps you solve the bottom one!

a. 5 + 2 = _____

35 + 2 = _____

b. 4 + 5 = _____

64 + 5 = _____

c. 3 + 6 = _____

93 + 6 = _____

3. Add. Below each problem, write a "helping" problem with numbers less than 10.

a. 52 + 7 = _____

2 + 7 = _____

b. 33 + 1 = _____

___ + ___ = _____

c. 11 + 5 = _____

___ + ___ = _____

4. The numbers are written in boxes! Add the ones in their own column.
   Copy the number of tens to the answer line.

a. $35 + 3$	b. $12 + 6$	c. $57 + 1$	d. $64 + 3$
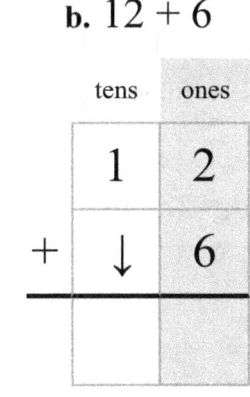			

	tens	ones
	3	5
+	↓	3
	3	*8*

	tens	ones
	1	2
+	↓	6

	tens	ones
	5	7
+	↓	1

	tens	ones
	6	4
+	↓	3

5. Now *you* write the numbers in the boxes. Add the ones in their own column.

a. $26 + 3$	b. $72 + 4$	c. $65 + 4$	d. $81 + 4$

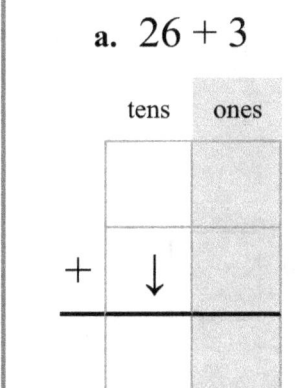 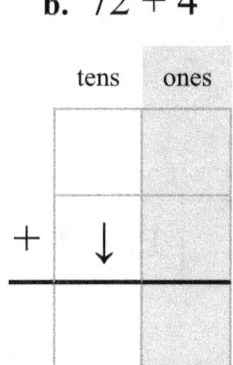 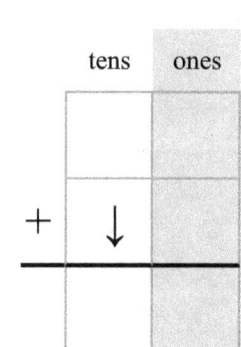 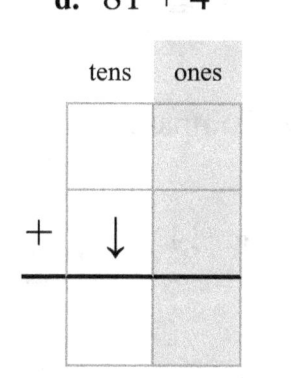

6. Add. Compare the problems.

a.	b.	c.	d.
$6 + 2 =$ _____	$4 + 3 =$ _____	$5 + 4 =$ _____	$11 + 7 =$ _____
$16 + 2 =$ _____	$24 + 3 =$ _____	$45 + 4 =$ _____	$61 + 7 =$ _____
$36 + 2 =$ _____	$34 + 3 =$ _____	$65 + 4 =$ _____	$41 + 7 =$ _____

7. Add three numbers.

a.	b.	c.
$20 + 5 + 2 =$ _____	$93 + 1 + 5 =$ _____	$100 + 5 + 4 =$ _____
$44 + 2 + 2 =$ _____	$83 + 4 + 3 =$ _____	$52 + 4 + 2 =$ _____

| Do you remember how to separate ("break up") a number into its TENS and ONES? | 
23 = 20 + 3
tens   ones | 
47 = 40 + 7
tens   ones |

8. Break the numbers into tens and ones or combine the tens and ones into numbers.

a.	b.	c.
18 = 10 + 8	32 = ____ + ____	_____ = 60 + 6
25 = ____ + ____	95 = ____ + ____	_____ = 9 + 80
55 = ____ + ____	____ = 40 + 9	_____ = 8 + 70

9. Compare. Write < , > , or =.

a. 24 + 3 ☐ 24 + 5	**c.** 17 + 2 ☐ 19 + 2	**e.** 58 ☐ 8 + 51
b. 83 + 5 ☐ 85 + 3	**d.** 36 + 4 ☐ 46 + 4	**f.** 66 ☐ 5 + 61

Puzzle Corner ☆ is a number we don't know—a mystery number! Your task is to *compare* without knowing the mystery number! For example, which is more, ☆ + 2 or ☆ + 7?

Write < or > in the boxes. Note: there is one comparison you **cannot** do without knowing the mystery number. Can you find it?

☆ + 5 ☐ ☆ + 4   ☆ − 5 ☐ ☆ − 4   ☆ − 5 ☐ ☆

☆ + 2 ☐ ☆ + 7   ☆ − 5 ☐ ☆ − 6   ☆ + ☆ ☐ ☆ + 20

Subtracting Without Borrowing

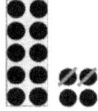 $14 - 2 = \underline{12}$ "I can subtract $4 - 2 = 2$; the 10 stays the same."	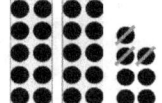 $27 - 3 = \underline{24}$ "I can subtract $7 - 3 = 4$; the 20 stays the same."	Think of the *ones digits* only. The tens do not change because we don't have to subtract from the tens.

1. Subtract and compare. The top problem helps you solve the bottom one!

a. $8 - 2 = \underline{\ 6\ }$ $28 - 2 = \underline{\ 26\ }$	**b.** $7 - 6 = \underline{\qquad}$ $17 - 6 = \underline{\qquad}$	**c.** $7 - 7 = \underline{\qquad}$ $67 - 7 = \underline{\qquad}$
d. $6 - 6 = \underline{\qquad}$ $56 - 6 = \underline{\qquad}$	**e.** $9 - 8 = \underline{\qquad}$ $49 - 8 = \underline{\qquad}$	**f.** $5 - 2 = \underline{\qquad}$ $95 - 2 = \underline{\qquad}$

2. Subtract. Write a "helping problem" below that uses only numbers less than 10.

a. $54 - 2 = \underline{\qquad}$ $4 - 2 = \underline{\qquad}$	**b.** $76 - 2 = \underline{\qquad}$ $\underline{\qquad} - \underline{\qquad} = \underline{\qquad}$	**c.** $88 - 4 = \underline{\qquad}$ $\underline{\qquad} - \underline{\qquad} = \underline{\qquad}$

3. Subtract. Cross out dots. Each box marked with a "T" stands for a ten.

a. $\begin{array}{cc} T & T \\ T & \end{array}$ $35 - 4 = \underline{\qquad}$ $35 - 3 = \underline{\qquad}$ $35 - 2 = \underline{\qquad}$	**b.** $\begin{array}{ccc} T & T & T \\ T & T & \end{array}$ $57 - 7 = \underline{\qquad}$ $57 - 5 = \underline{\qquad}$ $57 - 3 = \underline{\qquad}$	**c.** $\begin{array}{cc} T & T \\ T & T \end{array}$ $48 - 2 = \underline{\qquad}$ $48 - 4 = \underline{\qquad}$ $48 - 6 = \underline{\qquad}$	**d.** $\begin{array}{cc} T & T \\ T & \end{array}$ $34 - 1 = \underline{\qquad}$ $34 - 2 = \underline{\qquad}$ $34 - 4 = \underline{\qquad}$

4. Subtract.

a.	b.	c.	d.
77 – 6 = _____	47 – 2 = _____	57 – 4 = _____	15 – 3 = _____
22 – 1 = _____	75 – 1 = _____	86 – 2 = _____	98 – 4 = _____

5. Find the missing addends.

a. 10 + _____ = 15	**b.** 21 + _____ = 22	**c.** 65 + _____ = 69
32 + _____ = 38	94 + _____ = 95	33 + _____ = 36
72 + _____ = 79	44 + _____ = 48	91 + _____ = 98

6. Solve.

a. In the morning Katherine sold 21 pictures that she had painted, and in the afternoon she sold 7. How many pictures did she sell in total?

b. She had 30 pictures to sell when she started.
How many does she have left now?

c. Katherine can paint a picture in one hour. She started painting at 4:30 and painted three pictures. What time did she stop painting?

7. Take away all the ones (the dots) so that only the whole tens are left.

a. 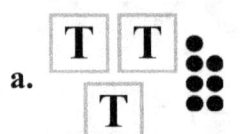	b.	c.
37 – _____ = 30	46 – _____ = 40	28 – _____ = _____ 
d. 57 – _____ = _____	e. 85 – _____ = _____	f. 69 – _____ = _____

8. Solve. In the last row, make your own problems, and let a friend solve them!

a. $50 + \bigcirc = 57$	**b.** $\bigcirc + 2 = 88$	**c.** $79 - 9 = \bigcirc$
d. $\bigcirc - 5 = 20$	**e.** $90 - \bigcirc = 85$	**f.** $42 = 40 + \bigcirc$
$\bigcirc + \underline{\quad} = \underline{\quad}$		$\underline{\quad} + \bigcirc = \underline{\quad}$

9. Count by fives. Notice the patterns! A 100-chart or an abacus can help you.

a. 10, 15, _____, _____, _____, _____, _____, _____, _____

b. 1, 6, _____, _____, _____, _____, _____, _____, _____

c. 3, 8, _____, _____, _____, _____, _____, _____, _____

10. Continue the patterns.

a.	b.	c.
$88 - 0 = \underline{\quad}$	$95 - 2 = \underline{\quad}$	$48 - 1 = \underline{\quad}$
$88 - 1 = \underline{\quad}$	$85 - 2 = \underline{\quad}$	$46 - 1 = \underline{\quad}$
$88 - 2 = \underline{\quad}$	$75 - 2 = \underline{\quad}$	$44 - 1 = \underline{\quad}$
$88 - \underline{\quad} = \underline{\quad}$	$\underline{\quad} - \underline{\quad} = \underline{\quad}$	$\underline{\quad} - 1 = \underline{\quad}$
$88 - \underline{\quad} = \underline{\quad}$	$\underline{\quad} - \underline{\quad} = \underline{\quad}$	$\underline{\quad} - \underline{\quad} = \underline{\quad}$
$\underline{\quad} - \underline{\quad} = \underline{\quad}$	$\underline{\quad} - \underline{\quad} = \underline{\quad}$	$\underline{\quad} - \underline{\quad} = \underline{\quad}$
$\underline{\quad} - \underline{\quad} = \underline{\quad}$	$\underline{\quad} - \underline{\quad} = \underline{\quad}$	$\underline{\quad} - \underline{\quad} = \underline{\quad}$

Adding or Subtracting Two-Digit Numbers

1. Subtract. Cover with your fingers, or cross out, what needs to be subtracted.

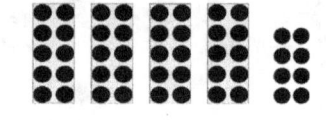	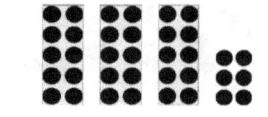	
a. $48 - 20 =$ _____	b. $36 - 30 =$ _____	c. $61 - 50 =$ _____
	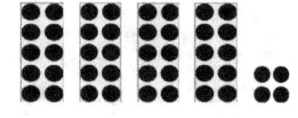	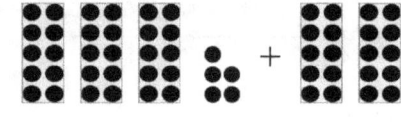
d. $55 - 40 =$ _____	e. $44 - 30 =$ _____	f. $72 - 50 =$ _____

2. Add. You can also use an abacus to help you, instead of the pictures.

	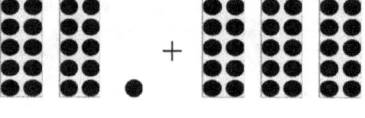	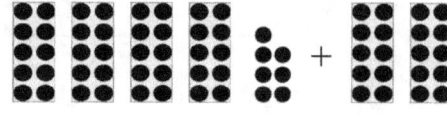
a. $35 + 20 =$ _____	b. $21 + 30 =$ _____	c. $47 + 20 =$ _____
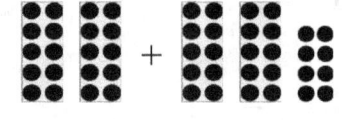	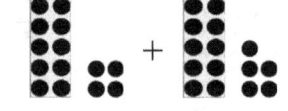	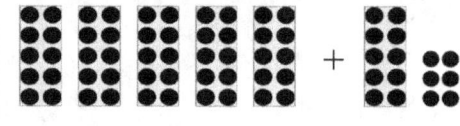
d. $20 + 28 =$ _____	e. $14 + 15 =$ _____	f. $50 + 16 =$ _____

3. A challenge! These might be a little difficult, so you can use a 100-bead abacus if you need help.

a.	b.	c.
$35 + 20 =$ _____	$40 + 17 =$ _____	$33 - 20 =$ _____
$76 + 30 =$ _____	$30 + 33 =$ _____	$78 - 50 =$ _____
$22 + 50 =$ _____	$56 - 20 =$ _____	$99 - 40 =$ _____

We can write the numbers to be added or subtracted one under the other in the boxes.

Then we add or subtract the ones in their own column (marked orange).

Then we add or subtract the tens in their own column (marked green).

tens	ones
4	5
+ 2	3
6	8

tens	ones
7	8
− 5	6
2	2

Which numbers were added?
What is the answer?

Which numbers were subtracted?
What is the answer?

4. Add.

a.

tens	ones
4	2
+ 2	4

b.

tens	ones
5	3
+ 0	6

c.

tens	ones
2	5
+ 5	3

d.

tens	ones
3	5
+ 0	4

5. Subtract.

a.

tens	ones
9	5
− 2	0

b.

tens	ones
5	8
− 2	6

c.

tens	ones
2	5
− 0	3

d.

tens	ones
7	9
− 6	4

6. Write one number under the other in the boxes and add.

a. 17 + 21   **b.** 34 + 14   **c.** 51 + 7   **d.** 32 + 5

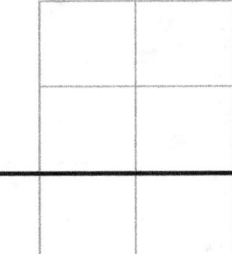
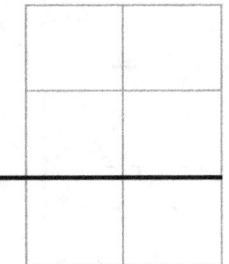
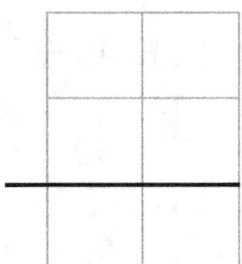
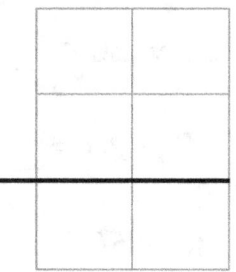

7. Subtract. Either use the picture or write the second number
   under the first number in the boxes.

3	8
− 1	4

a. 38 – 14 = _____

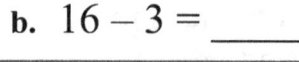

1	6
−	3

b. 16 – 3 = _____

c. 47 – 25 = _____

d. 38 – 26 = _____

8. The dots show the numbers to add. Write the numbers in the boxes. Add.

a. tens ones

2	4
1	3

b. tens ones

c. tens ones

d. tens ones

e. tens ones

f. tens ones

9. You should be able to do these problems in your head, without any help!

a. 30 + 20 = _____	**b.** 40 + 60 = _____	**c.** 60 – 40 = _____
60 + 20 = _____	30 + 30 = _____	70 – 50 = _____
30 + 50 = _____	50 – 20 = _____	90 – 40 = _____

10. Write the numbers in the boxes. Subtract the tens and the ones in their columns.

    a. 57 – 21   **b.** 74 – 14   **c.** 59 – 7   **d.** 99 – 58

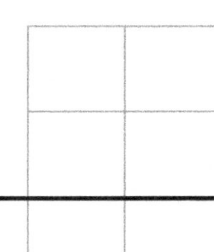 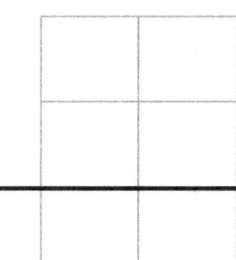 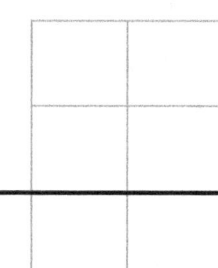

11. Solve the word problems. You can use an abacus to help or write the numbers one under the other in the boxes.

a. James went fishing and caught 28 fish. His wife cooked 11 fish for supper and put the rest into the freezer. How many fish did she put into the freezer?	
b. If you buy a shirt for $22 and jeans for $34, how much do you have to pay for both items?	
c. Mom is 38 years old and John is 11. How many years older is Mom than John?	
d. Jessica had 34 colored pencils and Matt had 22. Jessica gave Matt 6 pencils. Now how many does Matt have?	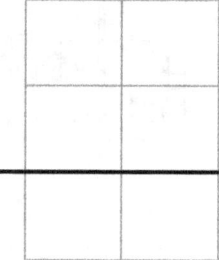

Completing the Next Ten

Review: What numbers make 10? You need to remember these well!	1 + _____ = 10	8 + _____ = 10	3 + _____ = 10
	7 + _____ = 10	5 + _____ = 10	6 + _____ = 10
	4 + _____ = 10	9 + _____ = 10	2 + _____ = 10

Completing the ten

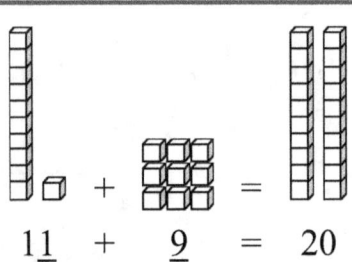

11 + 9 = 20

The 1 and the 9 single cubes make a *new* ten. We get a total of 20.

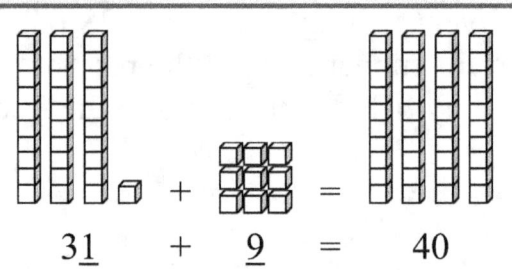

31 + 9 = 40

The 1 and the 9 single cubes make a *new* ten. We get a total of 40.

1. Draw more single blocks until there are ten of them. Circle the new ten.
   Write the addition that completes the next ten. You can also do these with an abacus.

a. 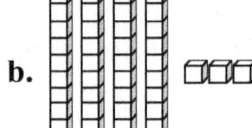 33 + _____ = _40_	**b.** 43 + _____ = _____	**c.** 27 + _____ = _____
d. 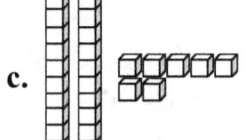 _____ + _____ = _____	**e.** _____ + _____ = _____	**f.** _____ + _____ = _____

2. Write the previous and next whole ten.

a. ___10___ , 13 , ___20___	b. _____ , 57 , _____	c. _____ , 46 , _____
d. _____ , 81 , _____	e. _____ , 78 , _____	f. _____ , 94 , _____

3. Draw more single blocks until there are ten of them. Circle the new ten. Write the
   addition that completes the next ten. You can also do these with an abacus.

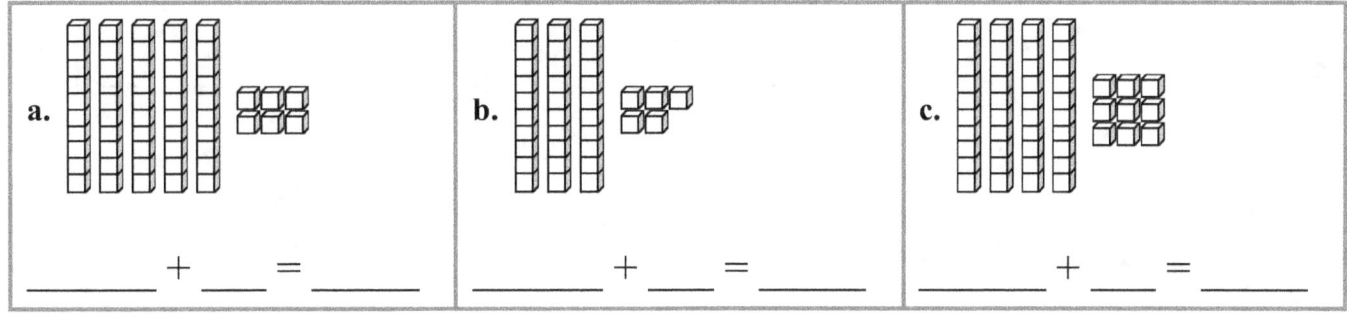

a. _____ + ___ = _____

b. _____ + ___ = _____

c. _____ + ___ = _____

4. Complete the next ten. The top problem is a "helping problem" for the bottom one.

a. 3 + ____ = 10	b. 4 + ____ = 10	c. 7 + ____ = 10
23 + ____ = 30	44 + ____ = ____	17 + ____ = ____

5. Complete the next ten. Think of the helping problem where you complete 10.

a. 13 + ( ) = 20	b. 21 + ( ) = 30	c. 74 + ( ) = 80
d. 88 + ( ) = 90	e. 44 + ( ) = 50	f. 96 + ( ) = 100
g. ( ) + 37 = 40	h. ( ) + 65 = 70	i. ( ) + 52 = 60
j. ( ) + 68 = 70	k. ( ) + 91 = 100	l. ( ) + 59 = 60

6. Complete the next ten. Then write a matching subtraction using the same numbers. Notice that the number in the oval is the same in both problems!

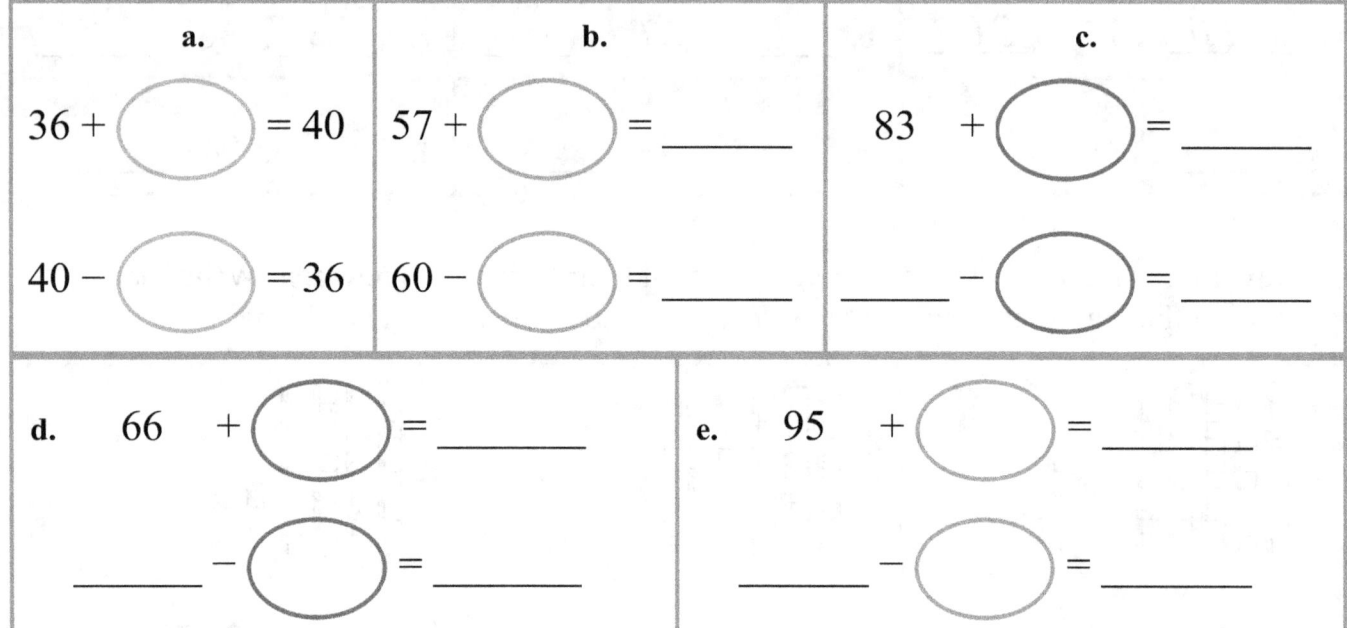

a.

$36 + \bigcirc = 40$

$40 - \bigcirc = 36$

b.

$57 + \bigcirc = \underline{\quad}$

$60 - \bigcirc = \underline{\quad}$

c.

$83 + \bigcirc = \underline{\quad}$

$\underline{\quad} - \bigcirc = \underline{\quad}$

d. $66 + \bigcirc = \underline{\quad}$

$\underline{\quad} - \bigcirc = \underline{\quad}$

e. $95 + \bigcirc = \underline{\quad}$

$\underline{\quad} - \bigcirc = \underline{\quad}$

7. A ticket to an amusement park costs $40. How much *more money* do these children need for the ticket?

a. Jeanine's mom gave her $30 for the ticket, and she has $7.

b. Derek already has $20. His parents will pay $10.

c. Muhammad has $12, and his mom has promised him $20.

Puzzle Corner   Find two different solutions to the puzzle.

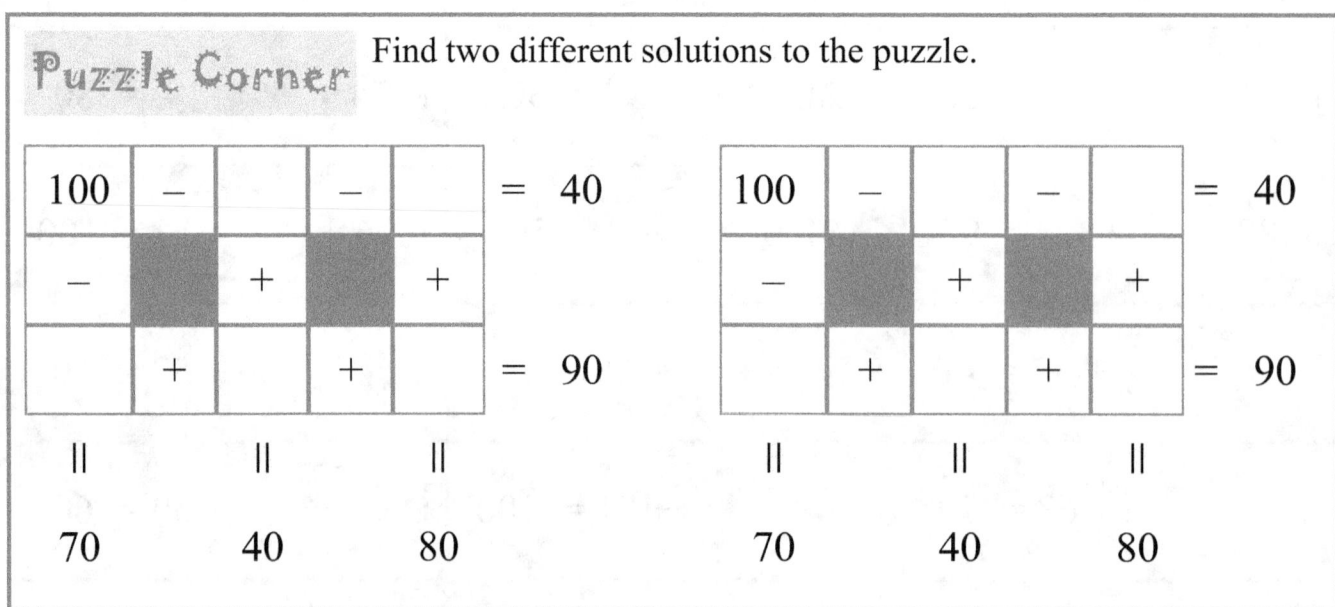

Going Over Ten

Remember? 10 plus 3, 4, 5, 6, 7, 8, or 9 makes one of the **TEEN** numbers!	10 plus <u>three</u> is <u>thir</u>teen. 10 plus <u>six</u> is <u>six</u>teen. 10 plus <u>nine</u> is <u>nine</u>teen. 10 plus <u>five</u> is <u>fif</u>teen.

1. Add. **a.** $10 + 4 =$ _____   **b.** $10 + 7 =$ _____

   c. $10 + 8 =$ _____   **d.** $10 + 3 =$ _____

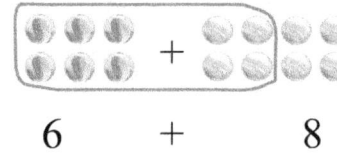

6 + 8

We circle TEN marbles to make a ten. We can now see that there are 10 and 4 marbles. $10 + 4 = 14$. So $6 + 8 = 14$.

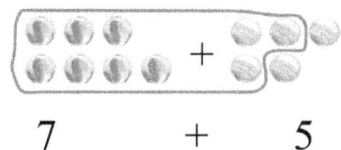

7 + 5

We circle TEN marbles to make a ten. We can now see that there are 10 and 2 marbles. $10 + 2 = 12$. So $7 + 5 = 12$.

2. First circle ten marbles to make a ten. How many marbles are there in all?

a. 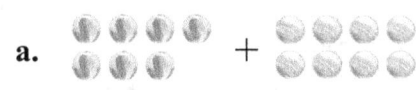
  7 + 8 = _____
	b.
	  8 + 8 = _____

c.
  6 + 5 = _____
	d.
	  9 + 4 = _____

e.
  8 + 5 = _____
	f.
	  8 + 9 = _____

g.
  7 + 7 = _____
	h.
	  9 + 9 = _____

Sums that go over to the next ten

Let's add 59 + 5. *First* we complete 60.

59 + 5

59 + 1 + 4

60 + 4 = 64

Because 59 + 1 = 60, we split the 5 into 1 and 4. The 1 goes with the 59 to make 60. Then the 60 and the 4 left over make 64.

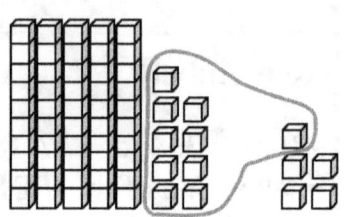

9 and 1 make a new ten. We get 6 tens.

59 + 5 = 64

3. Circle ten single cubes to make a ten. Count the tens and ones. Write the answer.

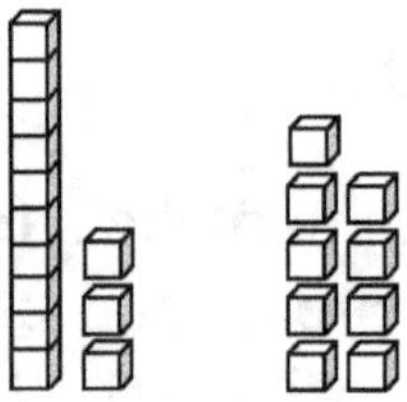

a. 13 + 9 = _____

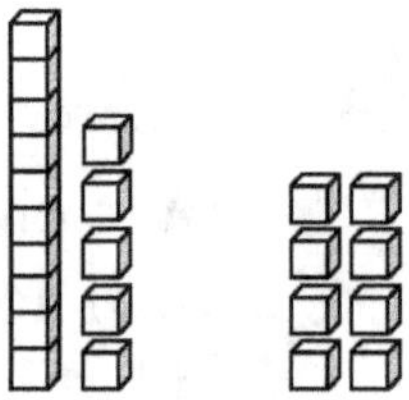

b. 15 + 8 = _____

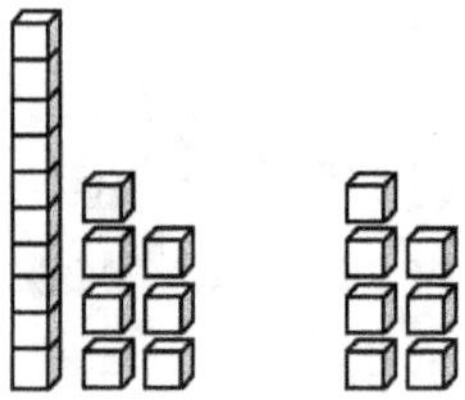

c. 17 + 7 = _____

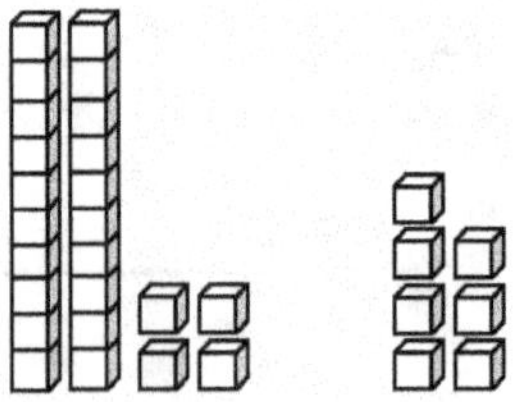

d. 24 + 7 = _____

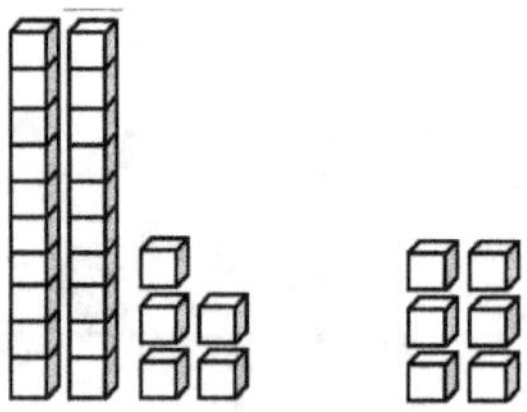

e. 25 + 6 = _____

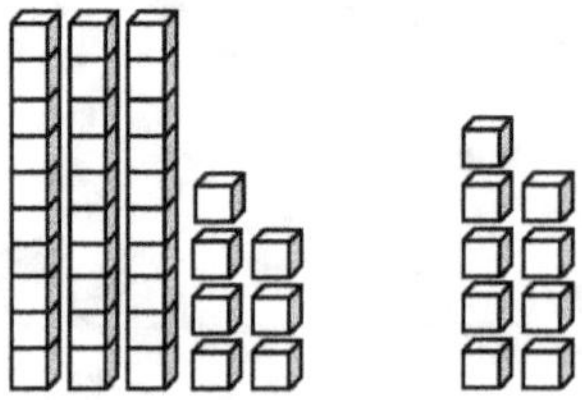

f. 37 + 9 = _____

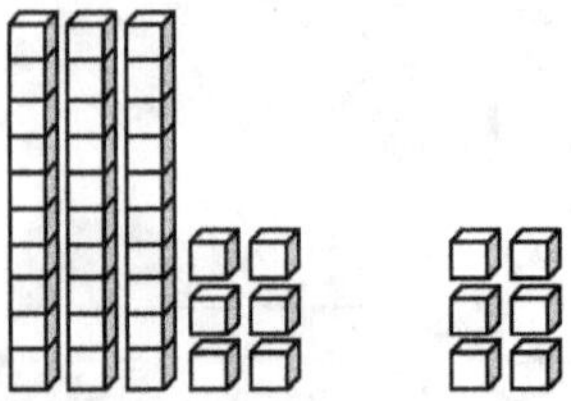

g. 36 + 6 = _____

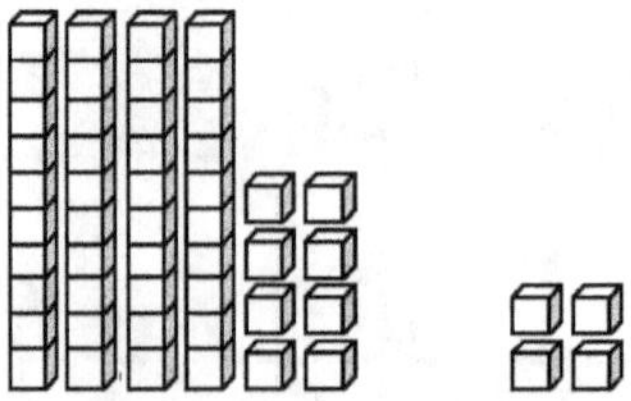

h. 48 + 4 = _____

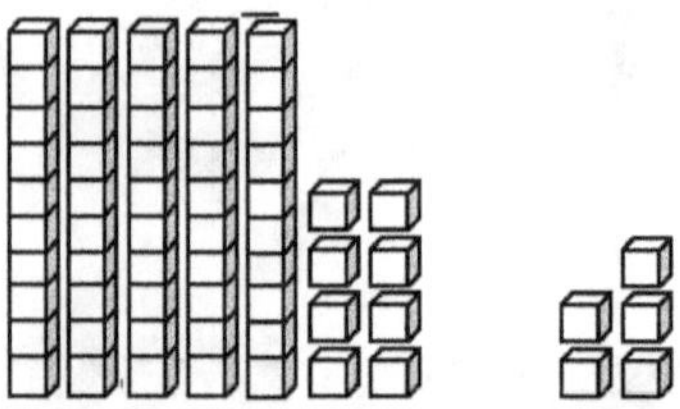

i. 58 + 5 = _____

4. Complete. Split the second number into two parts so that you can complete the next ten. You can also use your abacus to solve these.

a. 28 + 8	**b.** 47 + 5	**c.** 79 + 9
28 + _2_ + _6_	47 + _3_ + ____	79 + ____ + ____
30 + ____ = _____	50 + ____ = _____	80 + ____ = _____
d. 39 + 3	**e.** 27 + 5	**f.** 38 + 7
39 + ____ + ____	27 + ____ + ____	38 + ____ + ____
40 + ____ = _____	____ + ____ = _____	____ + ____ = _____

5. Add. First make a new ten with some of the single dots. You can also use your abacus.

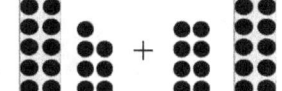

a. 25 + 15 = _____   b. 17 + 25 = _____   c. 38 + 26 = _____

d. 17 + 18 = _____   e. 14 + 48 = _____   f. 37 + 24 = _____

6. Add. Sometimes you can make a new ten and sometimes not. An abacus can help also.

a. 24 + 15 = _____   b. 18 + 22 = _____   c. 36 + 16 = _____

d. 15 + 23 = _____   e. 43 + 16 = _____   f. 25 + 37 = _____

7. The family counted how many birds they saw on their trip. They used tally marks.

		Count
Dad	卌 卌 卌 IIII	
Mom	卌 卌 卌 卌 卌 III	
Mary	卌 卌 II	
Mark	卌 卌 卌 卌 卌	
Angie	卌 卌 卌 卌 卌 卌 卌 I	

a. Fill in the *Count* column in the chart.

b. Make a bar graph (below).

c. How many more birds did Dad see than Mary?

d. How many more birds did Angie see than Mark?
   Use subtraction. Write the smaller number under the larger.

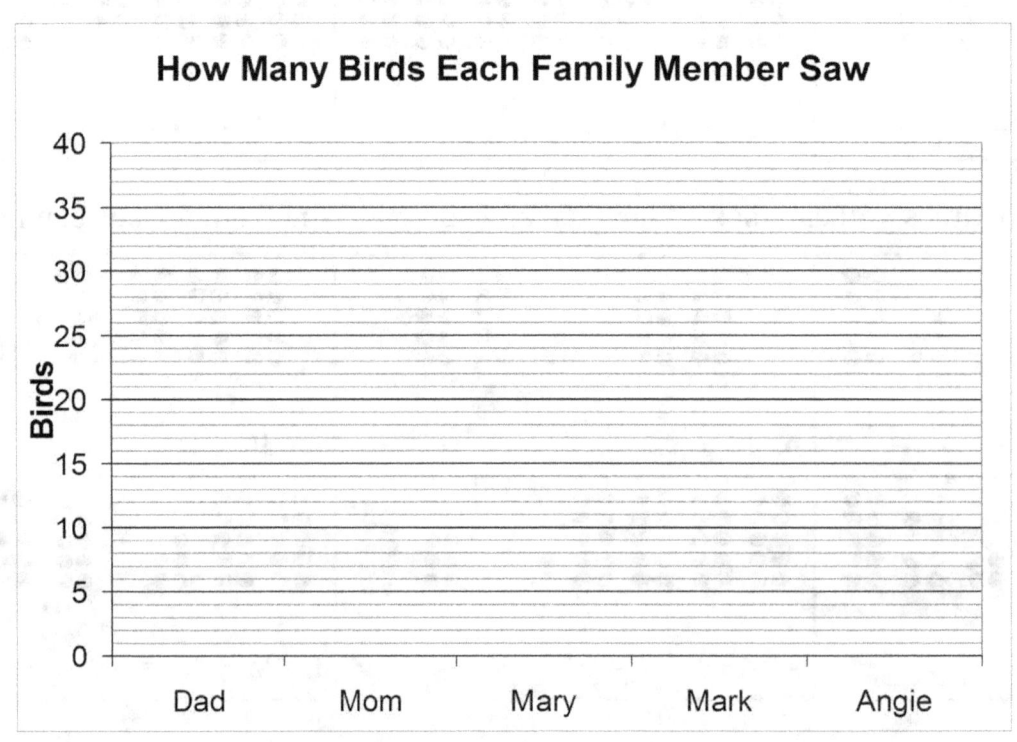

How Many Birds Each Family Member Saw

Birds

40
35
30
25
20
15
10
5
0

Dad   Mom   Mary   Mark   Angie

Subtracting from Whole Tens

One way to solve "30 – 7 = ?"
The picture shows 30 as two columns of ten and ten single blocks. Now you can easily subtract (by covering) seven blocks. 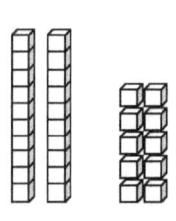 The answer is 23.

Another way to solve "30 – 7 = ?"
Think of the helping problem where you subtract from ten: 10 – 7 = 3. The answer must end in three. For the same reason, 30 – 7 must end in three in the previous ten, which is the twenties. So the answer must be 23.

1. Subtract. The last ten is broken into single blocks so you can cross some out!

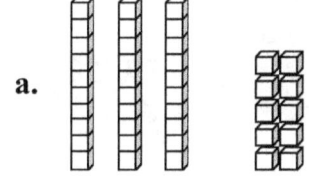 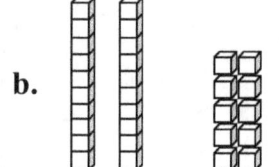 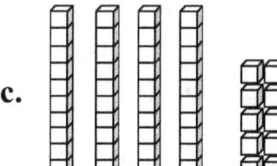 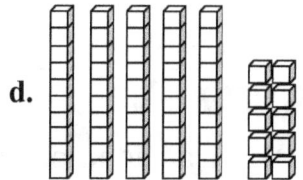

a.	b.	c.	d.
40 – 4 = _____	30 – 5 = _____	50 – 2 = _____	60 – 7 = _____
40 – 6 = _____	30 – 4 = _____	50 – 8 = _____	60 – 9 = _____
40 – 7 = _____	30 – 9 = _____	50 – 3 = _____	60 – 1 = _____
40 – 8 = _____	30 – 6 = _____	50 – 6 = _____	60 – 4 = _____

2. Subtract the same number three times.

a. 70 – 10 – 10 – 10 = _____	**b.** 90 – 20 – 20 – 20 = _____

3. Add and subtract whole tens.

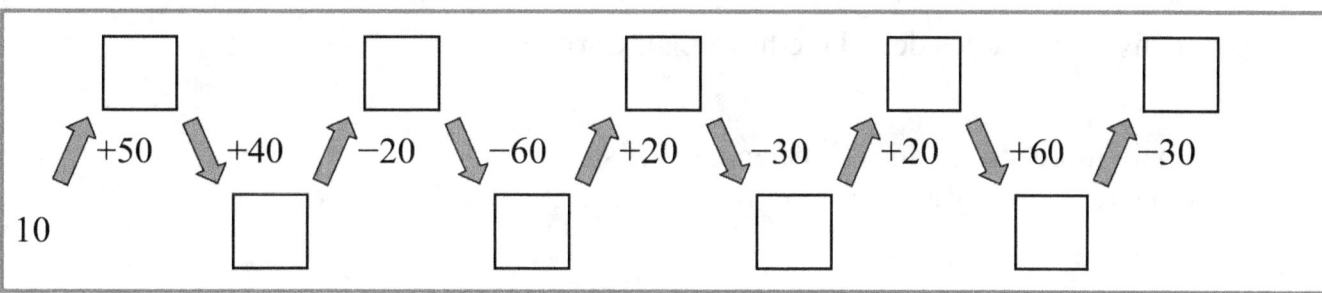

4. Subtract from whole tens.

a.	b.	c.	d.
$70 - 6 =$ _____	$50 - 8 =$ _____	$40 - 1 =$ _____	$100 - 5 =$ _____
$70 - 5 =$ _____	$50 - 7 =$ _____	$40 - 2 =$ _____	$100 - 7 =$ _____
$70 - 2 =$ _____	$50 - 6 =$ _____	$40 - 3 =$ _____	$100 - 9 =$ _____

5. Add and subtract the same number.

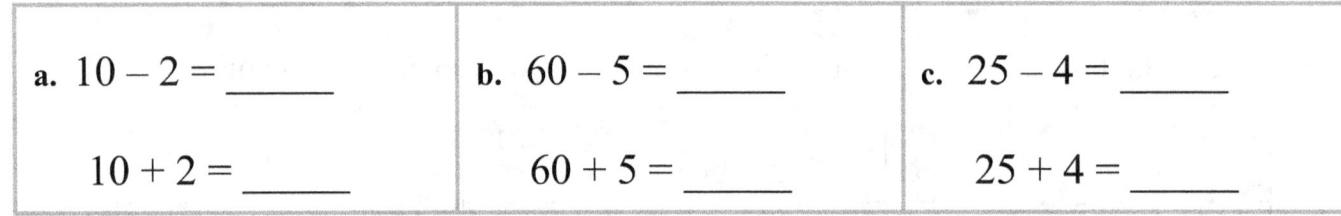

a. $10 - 2 =$ _____

$10 + 2 =$ _____

b. $60 - 5 =$ _____

$60 + 5 =$ _____

c. $25 - 4 =$ _____

$25 + 4 =$ _____

6. Write an addition or subtraction sentence for each problem and solve it.

a. There are 20 pupils in the class, and each one needed
a pencil. The teacher found only 16 pencils.
How many pencils do they still need?

b. There are 17 bushes growing in the front yard
and seven in the back yard. How many more are
in the front yard than in the back yard?

c. Carmen has 13 pretty stones, Jane has 18, and Julie has 20.

How many more stones does Julie have than Carmen?

How many more stones does Jane have than Carmen?

How many more stones does Carmen need if she
wants to have as many as Julie?

Add Using "Just One More"

Do you remember the numbers that add up to 10 ("the sums of 10")?
There are 9 and 1, and what others? List them now.

JUST ONE MORE than a sum of 10:			
$8 + \underline{2} = 10$ $8 + \underline{3} = 11$	8 + 3 is JUST ONE MORE than 8 + 2, so the answer is also just one more.	$\underline{5} + 5 = 10$ $\underline{6} + 5 = 11$	6 + 5 is JUST ONE MORE than 5 + 5, so the answer is also just one more.

1. Change the underlined number to be JUST ONE MORE. The answer changes, too!

a. $8 + \underline{2} = 10$ $8 + \underline{\ 3\ } = \underline{\hspace{1cm}}$	b. $4 + \underline{6} = 10$ $4 + \underline{\hspace{1cm}} = \underline{\hspace{1cm}}$	c. $\underline{7} + 3 = 10$ $\underline{\hspace{1cm}} + 3 = \underline{\hspace{1cm}}$
d. $\underline{1} + 9 = 10$ $\underline{\hspace{1cm}} + 9 = \underline{\hspace{1cm}}$	e. $5 + \underline{5} = 10$ $5 + \underline{\hspace{1cm}} = \underline{\hspace{1cm}}$	f. $\underline{4} + 4 = 8$ $\underline{\hspace{1cm}} + 4 = \underline{\hspace{1cm}}$

2. Find the missing numbers.

a. $7 + \boxed{} = 10$ $7 + \boxed{} = 11$	b. $8 + \boxed{} = 10$ $8 + \boxed{} = 11$	c. $6 + \boxed{} = 10$ $6 + \boxed{} = 11$
d. $5 + \boxed{} = 11$	e. $9 + \boxed{} = 11$	f. $3 + \boxed{} = 11$

3. Add. Think of JUST ONE MORE. Color the problems where you use that idea!

a.	b.	c.	d.
$7 + 2 = \underline{\hspace{1cm}}$	$5 + 6 = \underline{\hspace{1cm}}$	$4 + 6 = \underline{\hspace{1cm}}$	$2 + 9 = \underline{\hspace{1cm}}$
$3 + 8 = \underline{\hspace{1cm}}$	$3 + 4 = \underline{\hspace{1cm}}$	$2 + 8 = \underline{\hspace{1cm}}$	$5 + 4 = \underline{\hspace{1cm}}$
$5 + 5 = \underline{\hspace{1cm}}$	$6 + 4 = \underline{\hspace{1cm}}$	$7 + 4 = \underline{\hspace{1cm}}$	$3 + 7 = \underline{\hspace{1cm}}$

> The **double** of something means twice (two times) that thing.
> For example, "double four" means 4 and 4. So double 4 is 8.
> How much is double 3? Double 5?

> Double six, or 6 + 6, is 12.
> We can use that to find 6 + 7. It is JUST ONE MORE! It is 13.

4. On the right you see a **doubles chart**. You can use it for the addition
   problems below. Think of "JUST ONE MORE!"

a. $7 + 6 =$ ____		**b.** $7 + 7 =$ ____		**c.** $9 + 8 =$ ____
d. $8 + 8 =$ ____		**e.** $5 + 6 =$ ____		**f.** $9 + 10 =$ ____
g. $7 + 8 =$ ____		**h.** $9 + 9 =$ ____		**i.** $6 + 5 =$ ____
j. $8 + 9 =$ ____		**k.** $6 + 7 =$ ____		**l.** $8 + 7 =$ ____

$5 + 5 = 10$

$6 + 6 = 12$

$7 + 7 = 14$

$8 + 8 = 16$

$9 + 9 = 18$

5. Solve the word problems.

a. Joe bought a package of 12 balloons. He gave three to Sam, two to
his sister and five to Jane. How many balloons did he give away?

How many balloons does Joe have left?

b. Marsha found seven uniforms for the softball teams in one box, and six more
uniforms in another box. How many uniforms did Marsha find?

c. Three of the uniforms Marsha found were clean, but she had to wash the rest.
How many uniforms did Marsha have to wash?

d. Eight girls and five boys came to play softball.
How many more girls came than boys?

e. Did Marsha have enough uniforms for the boys and girls who came to
play softball?

If not, how many more uniforms does she need?

If so, how many uniforms were left over?

A "Trick" with Nine and Eight

A "trick" with nine

Imagine that nine wants to be ten! He is not happy—he wants to become a full TEN! So, nine asks the other number (this time, seven) to give him one in order to make him a ten.

Seven says, "Okay," gives one to nine, and has only six left for himself. In the end, we have 10 and 6. We get 16.

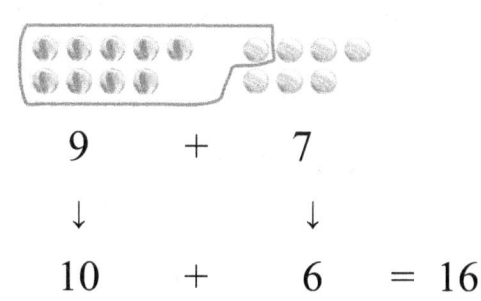

$$9 + 7$$
$$\downarrow \qquad \downarrow$$
$$10 + 6 = 16$$

We can also show the same thing this way →
Notice: it will also work if the second number is 9. Why? Because you can add in any order. 5 + 9 is the same as 9 + 5.

$$9 + 7$$
$$| \quad \backslash$$
$$9 + 1 + 6$$
$$10 + 6 = 16$$

1. Circle all of the blue marbles and enough of the yellow ones to make a ten. Add.

a. 9 + 6

$$10 + \underline{\ 5\ } = \underline{\qquad}$$

b. 9 + 4

$$10 + \underline{\qquad} = \underline{\qquad}$$

c. 9 + 3

$$10 + \underline{\qquad} = \underline{\qquad}$$

d. 9 + 5

$$10 + \underline{\qquad} = \underline{\qquad}$$

2. Fill in the blanks. Imagine that nine wants to become a ten.

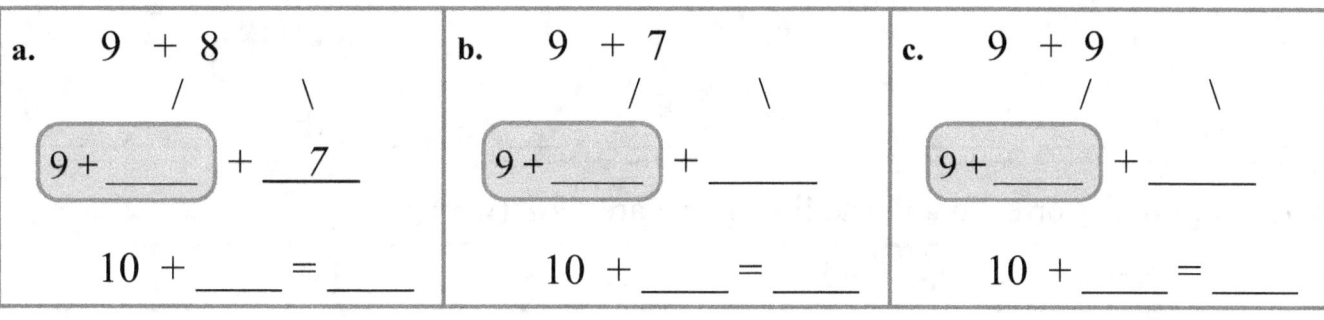

a. 9 + 8
$$/ \qquad \backslash$$
$$9 + \underline{\qquad} + \underline{\ 7\ }$$
$$10 + \underline{\qquad} = \underline{\qquad}$$

b. 9 + 7
$$/ \qquad \backslash$$
$$9 + \underline{\qquad} + \underline{\qquad}$$
$$10 + \underline{\qquad} = \underline{\qquad}$$

c. 9 + 9
$$/ \qquad \backslash$$
$$9 + \underline{\qquad} + \underline{\qquad}$$
$$10 + \underline{\qquad} = \underline{\qquad}$$

A "trick" with eight

Imagine that eight wants to be ten! She's not happy—she wants to become a full TEN! So eight asks the other number (this time, five) to give her two in order to make her a ten.

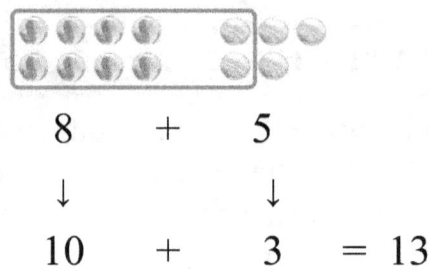

$$8 \quad + \quad 5$$

$$\downarrow \qquad \qquad \downarrow$$

Five says, "okay," gives two to eight, and has only three left for himself. In the end, we have 10 and 3. We get 13.

$$10 \quad + \quad 3 \quad = 13$$

We can also show the same thing this way:

$$8 + 5$$
$$| \quad \backslash$$
$$8 + 2 + 3$$
$$10 + 3 = 13$$

3. Circle all of the blue marbles and enough of the yellow ones to make a ten. Add.

a. 8 + 6

10 + _____ = _____

b. 8 + 7

10 + _____ = _____

c. 8 + 3

10 + _____ = _____

d. 8 + 4

10 + _____ = _____

4. Fill in the blanks. Imagine that eight wants to become a ten.

a. 8 + 8
/ \
8 + _2_ + ____
10 + ____ = ____

b. 8 + 5
/ \
8 + ____ + ____
10 + ____ = ____

c. 8 + 7
/ \
8 + ____ + ____
10 + ____ = ____

5. Right or not? Correct out the additions that are *false* (wrong).

a. $6 + 6 = 13$	**b.** $7 + 8 = 15$	**c.** $9 + 6 = 15$	**d.** $9 + 7 = 17$

6. Solve.

a. A basket had nine apples in it. Alice ate two, and her brother ate one. How many apples are left?	**b.** Jeremy picked up nine apples that were lying under an apple tree. He picked up six more apples that were lying under another tree. How many apples does Jeremy have now?
c. Alice picked 7 flowers and Jeremy picked 9. How many more flowers did Jeremy pick? How many flowers did the children have altogether?	**d.** Jeremy put three toy cars end-to-end. The first car was 5 cm long, the second was also 5 cm long, and the third car was 4 cm long. How long was Jeremy's line of cars?

7. Write a number inside the balloon so the numbers in the balloon make ten. Add.

a.	b.	c.
$7 + \underline{3} + 5 = \underline{15}$	$9 + \underline{} + 2 = \underline{}$	$7 + \underline{} + 5 = \underline{}$
d.	**e.**	**f.**
$6 + \underline{} + 6 = \underline{}$	$8 + \underline{} + 4 = \underline{}$	$5 + \underline{} + 8 = \underline{}$

8. Add. Think how the nine or the eight wants to be ten! If the *second* number is 8 or 9, turn the addition around. You can add the numbers in the other order, 8 or 9 first.

a. $8 + 6 =$ _____ b. $6 + 9 =$ _____ c. $9 + 4 =$ _____

d. $4 + 8 =$ _____ e. $8 + 7 =$ _____ f. $9 + 9 =$ _____

g. $9 + 5 =$ _____ h. $8 + 8 =$ _____ i. $3 + 8 =$ _____

Which numbers go in the triangles?   *Puzzle Corner*

a. $\triangle + 8 = 16$ b. $\triangle + 9 = 15$ c. $\triangle + 2 + 7 = 13$

Adding Within 20

You have learned several ways to help you add when the sum (the answer) is more than 10. Let's review them:

1. The trick with nine and eight.

$9 + 6 = ?$

Think of nine wanting to be ten, so six gives one to nine. Then the addition becomes $10 + 5$, which is 15.

2. Just one more than an addition you know.

For example, $3 + 7 = 10$, so $3 + 8$ must be just one more, or 11.

3. The doubles chart:

$5 + 5 = 10$
$6 + 6 = 12$
$7 + 7 = 14$
$8 + 8 = 16$
$9 + 9 = 18$

4. Just one more than a double:

$7 + 8$ is just one more than $7 + 7$.

Since $7 + 7$ is 14, then $7 + 8$ must be 15.

1. Finish these additions using the idea of "just one more" than a double.

a. $5 + 5 = 10$	**b.** $6 + 6 = 12$	**c.** $7 + 7 = 14$
$\underline{\ 5\ } + \underline{\ 6\ } = 11$ and	____ + ____ = 13 and	____ + ____ = _____ and
$\underline{\ 6\ } + \underline{\ 5\ } = 11$	____ + ____ = 13	____ + ____ = _____
d. $8 + 8 = 16$	**e.** $9 + 9 = 18$	**f.** $10 + 10 = 20$
____ + ____ = 17 and	____ + ____ = 19 and	____ + ____ = _____ and
____ + ____ = 17	____ + ____ = _____	____ + ____ = _____

2. Add. Use the trick with nine.

a. $9 + 8 =$ ____	**b.** $3 + 9 =$ ____	**c.** $9 + 5 =$ ____	**d.** $6 + 9 =$ ____

3. For each sum with 10 write another that is "just one more."

a. $1 + 9 = 10$	b. $3 + 7 = 10$	c. $8 + 2 = 10$
_____ + _____ = 11	_____ + _____ = 11	_____ + _____ = 11
d. $6 + 4 = 10$	e. $5 + 5 = 10$	f. $7 + 3 = 10$
_____ + _____ = 11	_____ + _____ = 11	_____ + _____ = 11

4. Add. Tell which idea you use to add.

Trick with nine	a. $7 + 7 = $ _____	b. $9 + 7 = $ _____	Doubles chart
Trick with eight	c. $8 + 3 = $ _____	d. $6 + 7 = $ _____	Just one more than a double
"Just one more" than a sum with 10	e. $5 + 6 = $ _____	f. $5 + 8 = $ _____	I just know it!
	g. $8 + 8 = $ _____	h. $4 + 9 = $ _____	

5. Solve.

a. Maria had $9. Then her mom gave her $5 for picking berries.
   Then she bought ice cream for $2.
   How much does Maria have now?

b. Ashley had 9 shirts and her brother Andy had 8. Then they both got
   three new shirts from their aunt. Now, who has more shirts?
   How many more?

c. Emily had $10. She bought colored pencils for $6 and a pretty
   eraser for $1. Now how much money does she have?

d. Natalie and Eric went to play tennis. They had 8 tennis balls with them.
   During the game they lost two balls, but they also found four more balls
   near the tennis court that other people had lost.
   Now how many tennis balls do they have?

6. *Two* more is added each time to the previous problem. Can you see the patterns?

a. 8 + 2 = _____	b. 5 + 3 = _____	c. 9 + 2 = _____	d. 7 + 3 = _____
8 + 4 = _____	5 + 5 = _____	9 + 4 = _____	7 + 5 = _____
8 + 6 = _____	5 + 7 = _____	9 + 6 = _____	7 + 7 = _____
8 + 8 = _____	5 + 9 = _____	9 + 8 = _____	7 + 9 = _____

7. Add and subtract. Start with the number in the bottom left corner and follow the arrows.

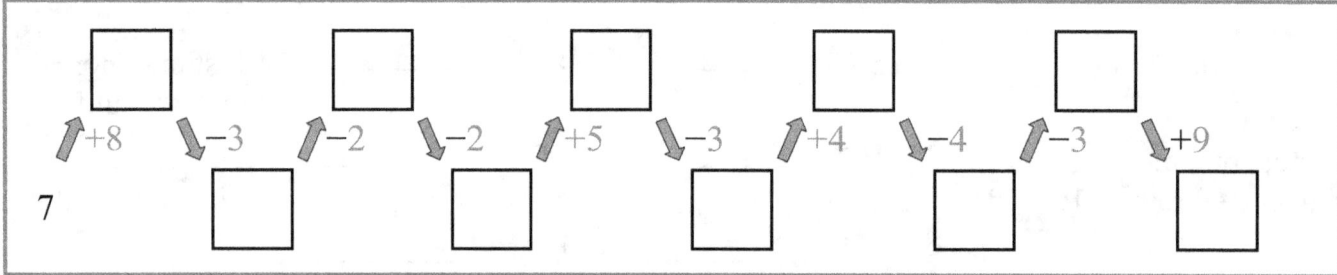

8. Count by tens.

a. 18, 28, _____, _____, _____, _____, _____, _____, _____

b. _____, _____, _____, _____, _____, 77, 87, _____, _____

9. What number goes in the triangle?

a. 6 + △ = 12

6 + △ = 13

b. 8 + △ = 16

9 + △ = 16

c. 6 + △ = 11

7 + △ = 11

10. Erica drew a line 9 cm long. Right after it, she drew another, 6 cm long. How long are her two lines together? You draw them, too!

11. *A challenge!* Here are ALL the addition facts where the sum is more than 10. How many of them can you solve? We will study these more in second grade.

a.	b.	c.	d.
8 + 8 = _____	7 + 8 = _____	7 + 7 = _____	5 + 8 = _____
2 + 9 = _____	9 + 6 = _____	9 + 8 = _____	3 + 9 = _____
7 + 5 = _____	6 + 5 = _____	7 + 4 = _____	7 + 6 = _____
e.	**f.**	**g.**	**h.**
9 + 4 = _____	8 + 6 = _____	9 + 2 = _____	6 + 9 = _____
4 + 8 = _____	6 + 6 = _____	8 + 5 = _____	8 + 7 = _____
6 + 7 = _____	5 + 9 = _____	5 + 7 = _____	8 + 4 = _____
i.	**j.**	**k.**	**l.**
9 + 3 = _____	4 + 9 = _____	9 + 9 = _____	8 + 9 = _____
4 + 7 = _____	7 + 7 = _____	6 + 8 = _____	5 + 6 = _____
9 + 5 = _____	3 + 8 = _____	6 + 6 = _____	8 + 3 = _____

Puzzle Corner

What numbers can go into these puzzles?

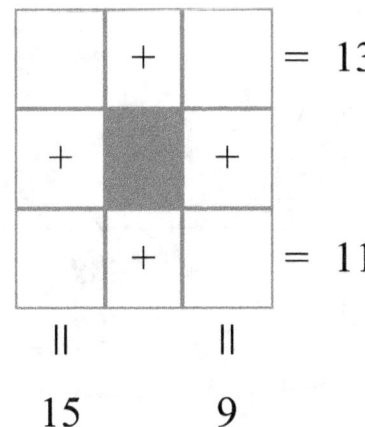

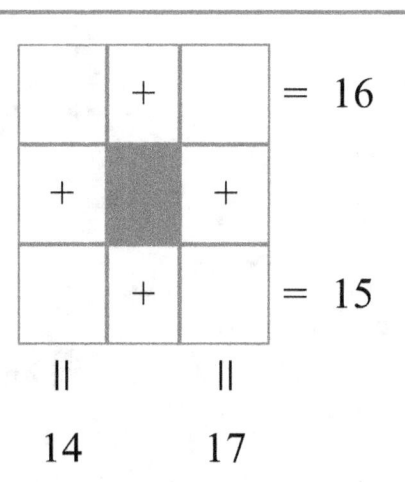

119

Subtract to 10

1. Subtract the "dots" that are not in the group of ten. You should only have ten left!

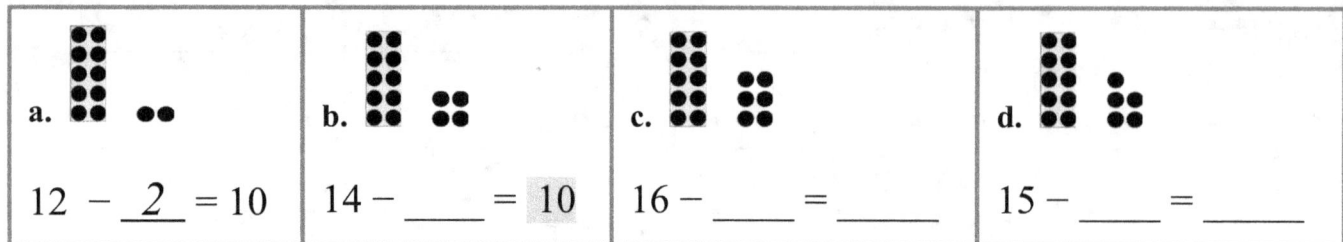

a. $12 - \underline{\ 2\ } = 10$

b. $14 - \underline{\quad} = 10$

c. $16 - \underline{\quad} = \underline{\qquad}$

d. $15 - \underline{\quad} = \underline{\qquad}$

2. Subtract the "ones" so that 10 is left.

a. $13 - \underline{\qquad} = 10$

b. $17 - \underline{\qquad} = \underline{\qquad}$

c. $19 - \underline{\qquad} = \underline{\qquad}$

Subtracting in parts

Let's subtract $13 - 5$. First we subtract enough dots that we have only 10 left. So we take away 3 dots. $13 - 3 = 10$.

We still need to subtract 2 more. We subtract those from 10. There are 8 left.

$$13 - 5$$
$$/ \ \backslash$$
$$13 - 3 - 2$$
$$= \underline{\ 8\ }$$

---

3. First subtract enough dots so that you have only 10 left. Then subtract the rest.

a. $14 - 7$
  / \
  $14 - 4 - 3$
  $= \underline{\qquad}$

b. $15 - 8$
  / \
  $15 - \underline{\quad} - \underline{\quad}$
  $= \underline{\qquad}$

c. $16 - 8$
  / \
  $16 - \underline{\quad} - \underline{\quad}$
  $= \underline{\qquad}$

d. $13 - 6$
  / \
  $13 - \underline{\quad} - \underline{\quad}$
  $= \underline{\qquad}$

e. $12 - 6$
  / \
  $12 - \underline{\quad} - \underline{\quad}$
  $= \underline{\qquad}$

f. $13 - 4$
  / \
  $13 - \underline{\quad} - \underline{\quad}$
  $= \underline{\qquad}$

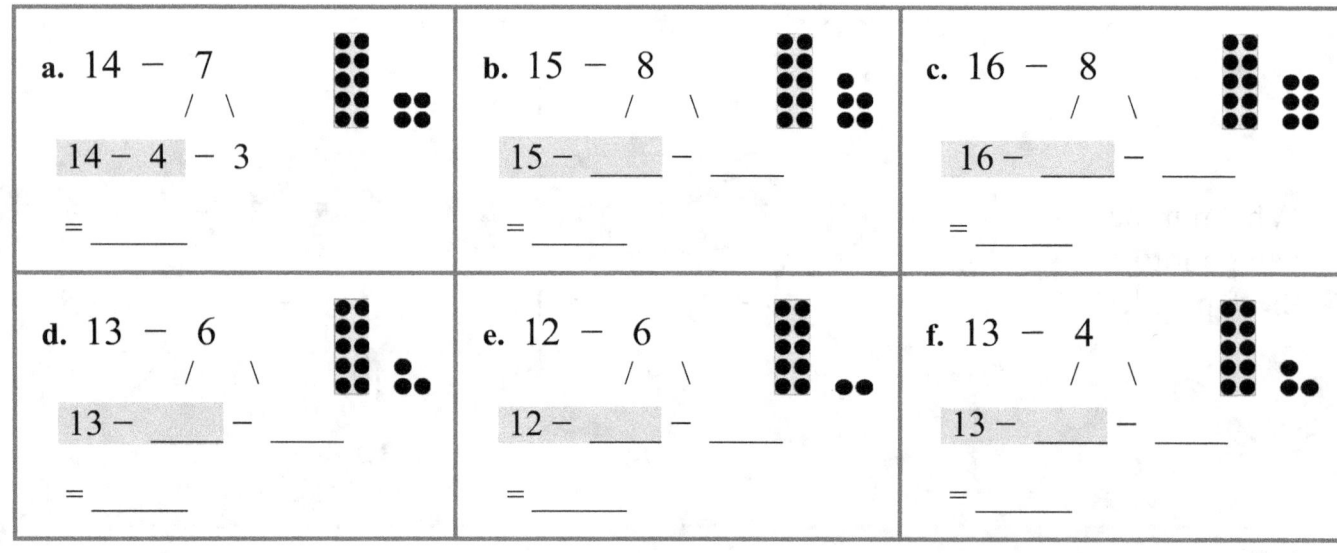

4. First subtract to 10. Then subtract the rest.

a. $12 - 6$ / \\ $12 - \underline{2} - 4$ $= \underline{\hspace{1cm}}$	**b.** $15 - 9$ / \\ $15 - \underline{\hspace{0.6cm}} - \underline{\hspace{0.6cm}}$ $= \underline{\hspace{1cm}}$	**c.** $13 - 8$ / \\ $13 - \underline{\hspace{0.6cm}} - \underline{\hspace{0.6cm}}$ $= \underline{\hspace{1cm}}$
d. $13 - 7$ / \\ $13 - \underline{\hspace{0.6cm}} - \underline{\hspace{0.6cm}}$ $= \underline{\hspace{1cm}}$	**e.** $14 - 7$ / \\ $14 - \underline{\hspace{0.6cm}} - \underline{\hspace{0.6cm}}$ $= \underline{\hspace{1cm}}$	**f.** $12 - 4$ / \\ $12 - \underline{\hspace{0.6cm}} - \underline{\hspace{0.6cm}}$ $= \underline{\hspace{1cm}}$

5. First subtract the dots that are not in the group of ten.

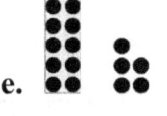

a. $12 - 5 = \underline{\hspace{1cm}}$	**b.** $14 - 6 = \underline{\hspace{1cm}}$	**c.** $13 - 6 = \underline{\hspace{1cm}}$	**d.** $15 - 7 = \underline{\hspace{1cm}}$
e. $15 - 8 = \underline{\hspace{1cm}}$	**f.** $14 - 5 = \underline{\hspace{1cm}}$	**g.** $16 - 8 = \underline{\hspace{1cm}}$	**h.** $13 - 8 = \underline{\hspace{1cm}}$

6. Tom is 13, Juan is 8, and Alice is 9.

   a. How many years older is Tom than Juan?

   b. How many years older is Tom than Alice?

   c. Two from now, how many years older than Juan will Tom be?

7. Finish this addition and subtraction "journey"!

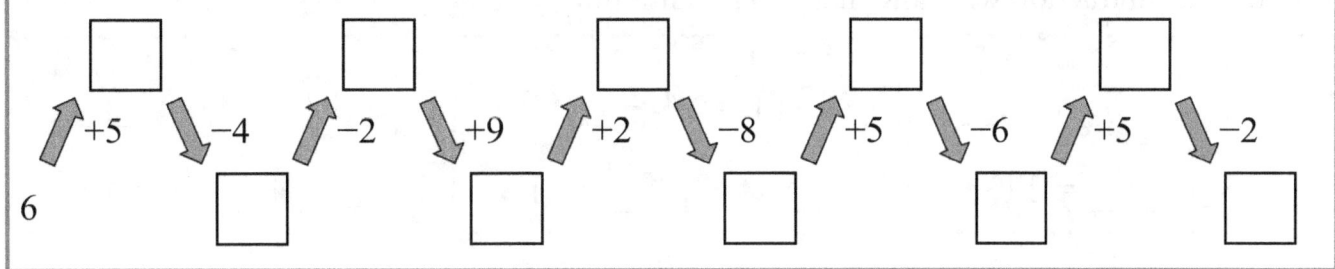

Using Addition to Subtract

From the picture on the right we can write two additions and two subtractions: a **fact family**.

There are TWO parts that make up the total, 6 and 5. In addition, we add the parts and get the total.

In subtraction, we start with the total and take away one of the parts. What is left? The other part.

Fact family with 6, 5, and 11:

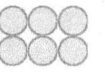 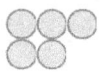

$6 + 5 = 11$	$11 - 5 = 6$
$5 + 6 = 11$	$11 - 6 = 5$

1. Write fact families.

a.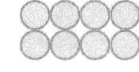

_____ + _____ = _____

_____ + _____ = _____

_____ − _____ = _____

_____ − _____ = _____

b.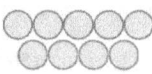

_____ + _____ = _____

_____ + _____ = _____

_____ − _____ = _____

_____ − _____ = _____

2. Calculate the total. Write two subtractions with the same numbers. Subtract from the total.

a. $8 + 4 =$ _____

_____ − _____ = _____

_____ − _____ = _____

b. $9 + 7 =$ _____

_____ − _____ = _____

_____ − _____ = _____

c. $7 + 6 =$ _____

_____ − _____ = _____

_____ − _____ = _____

3. For each subtraction write the matching addition.

a. $11 - 3 =$ ____

$3 +$ ____ $= 11$

b. $11 - 4 =$ ____

____ $+$ ____ $= 11$

c. $12 - 3 =$ ____

____ $+$ ____ $= 12$

When you have a subtraction problem like $15 - 9$ or $16 - 8$,
try to find the matching addition.

$15 - 9 = ?$	$16 - 8 = ?$
Think: $9 +$ ___ $= 15$	Think: $8 +$ ___ $= 16$
Or 9 and how many more is 15?	Or 8 and what number makes 16?
Guess and check!	Guess and check!
Will $9 + 8$ work? Or $9 + 7$? Or $9 + 6$? Or $9 + 5$? You can use the trick with nine!	Will $8 + 5$ work? Or $8 + 6$? Or $8 + 7$?

4. Solve each subtraction by thinking about the matching addition.

a. $14 - 8 =$ ___ $8 +$ ___ $= 14$	**b.** $15 - 7 =$ ___ $7 +$ ___ $= 15$	**c.** $17 - 8 =$ ___ $8 +$ ___ $= 17$
d. $12 - 8 =$ ___ ___ $+$ ___ $= 12$	**e.** $16 - 7 =$ ___ ___ $+$ ___ $= 16$	**f.** $13 - 7 =$ ___ ___ $+$ ___ $= 13$
g. $13 - 8 =$ ___ ___ $+$ ___ $= 13$	**h.** $11 - 7 =$ ___ ___ $+$ ___ $= 11$	**i.** $14 - 9 =$ ___ ___ $+$ ___ $= 14$

5. Doubles and doubles plus one more on the night sky! Solve. Also, match each
   addition to the subtraction in the same family.

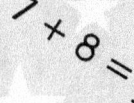
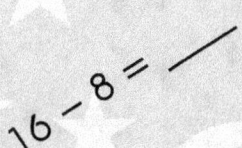
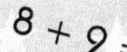

$8 + 8 =$ ___

$7 + 8 =$ ___

$17 - 8 =$ ___

$15 - 7 =$ ___

$16 - 8 =$ ___

$7 + 7 =$ ___

$8 + 9 =$ ___

$14 - 7 =$ ___

6. Now try to solve these subtractions by thinking of addition!

a. $12 - 8 =$ _____	b. $11 - 7 =$ _____	c. $13 - 9 =$ _____
d. $15 - 6 =$ _____	e. $18 - 9 =$ _____	f. $16 - 7 =$ _____

7. Solve.

a. Marsha had 15 crayons and Susana had 6. Marsha gave six of hers to Susana.
Now how many crayons does Marsha have?

And Susana?

Who has more?   How many more?

---

b. Judy counted seven stars in her drawing, and she thought, "That is not enough."
So she drew eight more. How many stars are in her drawing now?

---

c. Matthew had $12. He bought a book for $7.
Now how much money does he have left?

---

d. John bought a toy truck for $6 and a toy backhoe for $8.
The shopkeeper said, "That makes $15."
John said, "That's not right, it makes $13."

Who is right?

8. Connect the problems to the right answer.

$14 - 9$		$13 - 9$		$12 - 6$		$14 - 6$
	7				**9**	
$12 - 5$		$14 - 7$		$15 - 7$		$16 - 7$
	5				**6**	
$12 - 8$		$11 - 6$		$18 - 9$		$15 - 9$
	4				**8**	
$13 - 8$		$15 - 8$		$16 - 7$		$13 - 7$

Some Mixed Review

1. Write the time using numbers.

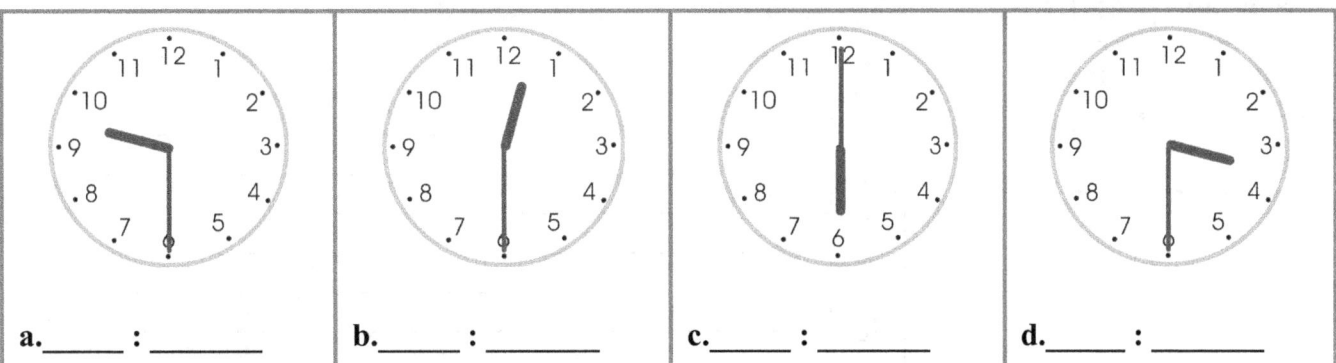

a._____ : _____   b._____ : _____   c._____ : _____   d._____ : _____

2. Write the time a half-hour later. Use numbers.

Now it is:	a. 2:00	b. 8:00	c. 12:00	d. 7:30	e. 10:30
A half-hour later, it is:					

3. Continue the patterns.

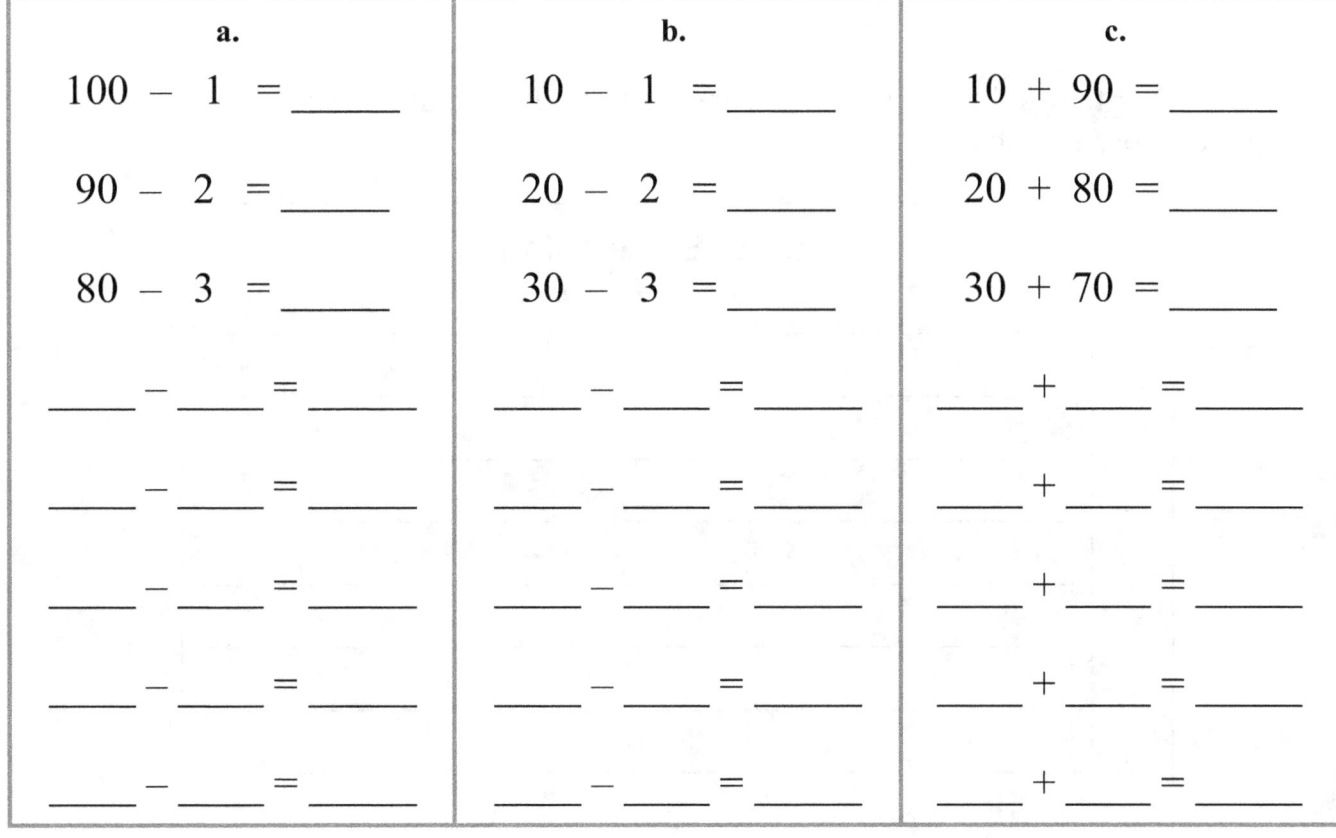

a.	b.	c.
100 – 1 = _____	10 – 1 = _____	10 + 90 = _____
90 – 2 = _____	20 – 2 = _____	20 + 80 = _____
80 – 3 = _____	30 – 3 = _____	30 + 70 = _____
____ – ____ = ____	____ – ____ = ____	____ + ____ = ____
____ – ____ = ____	____ – ____ = ____	____ + ____ = ____
____ – ____ = ____	____ – ____ = ____	____ + ____ = ____
____ – ____ = ____	____ – ____ = ____	____ + ____ = ____
____ – ____ = ____	____ – ____ = ____	____ + ____ = ____

4. Some children played a board game. Here are their scores. Draw a bar graph.

Maria	30
Santiago	45
Jeff	20
Elizabeth	35

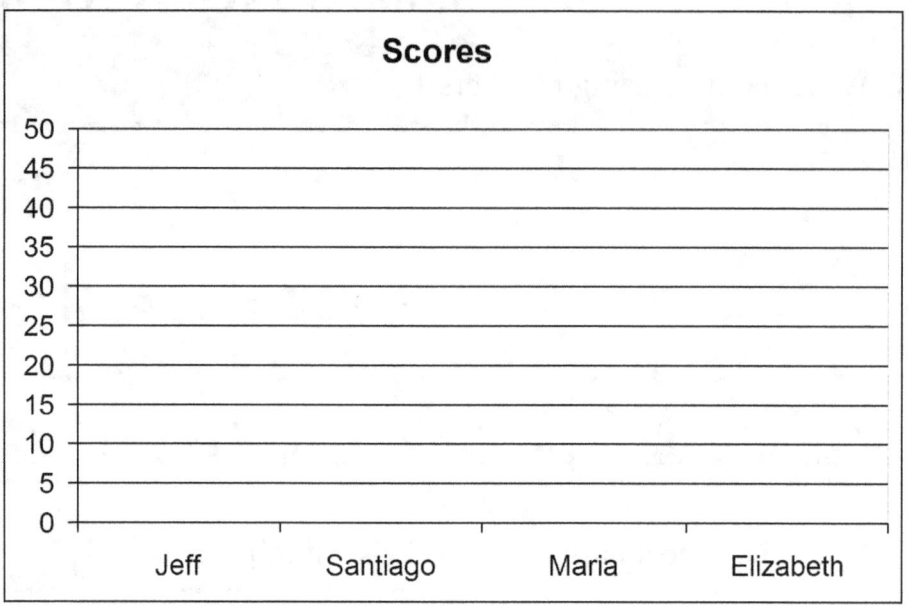

a. How many more points did Maria get than Jeff?

b. How many more points did Santiago get than Elizabeth?

5. Measure four things that are at most 12 inches long. Then make a bar graph about the length of those four things.

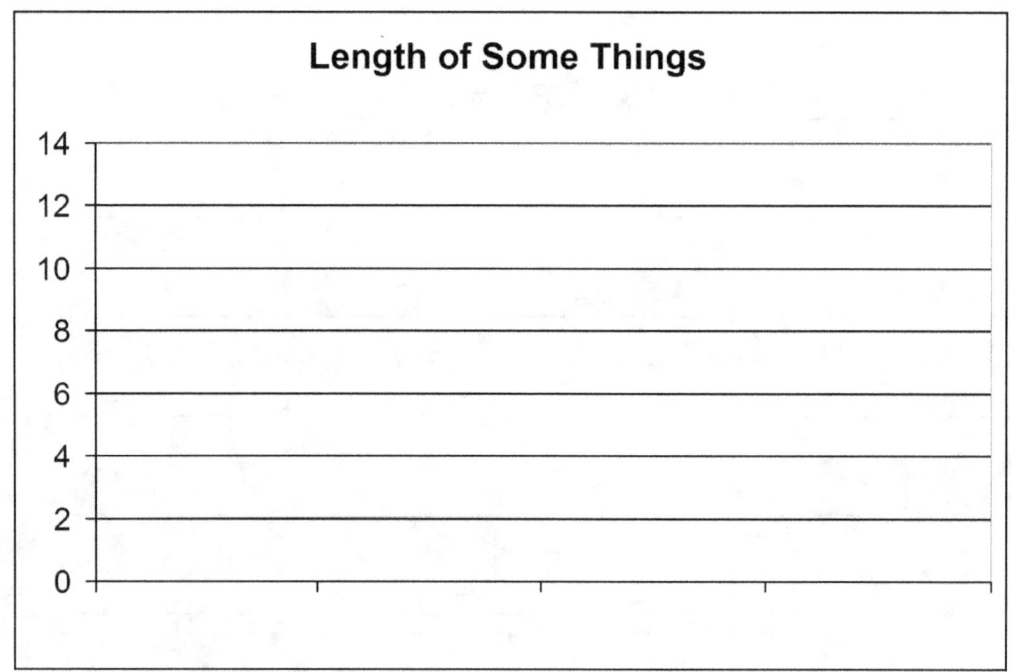

6. Below, draw some dots and connect them with lines so that you get a *triangle*.

7. Divide these shapes by drawing a straight line or lines from dot to dot. Then color as you are asked.

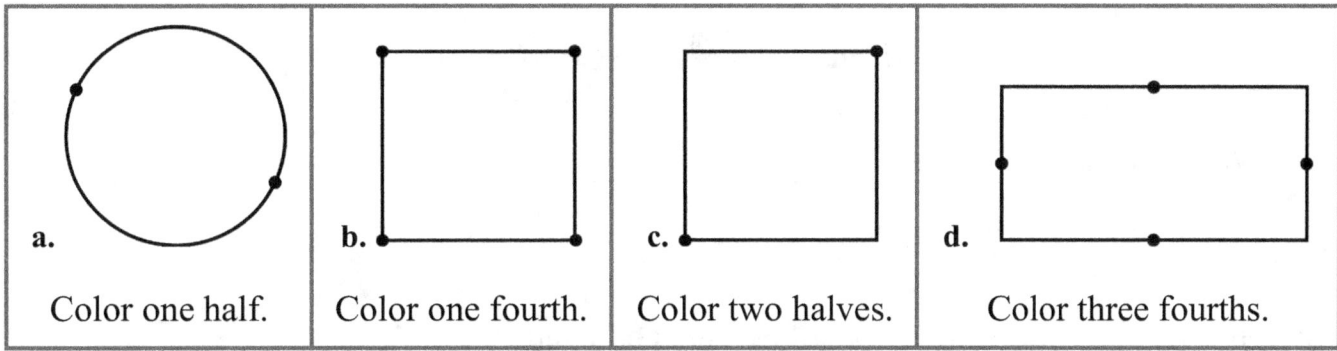

a.	b.	c.	d.
Color one half.	Color one fourth.	Color two halves.	Color three fourths.

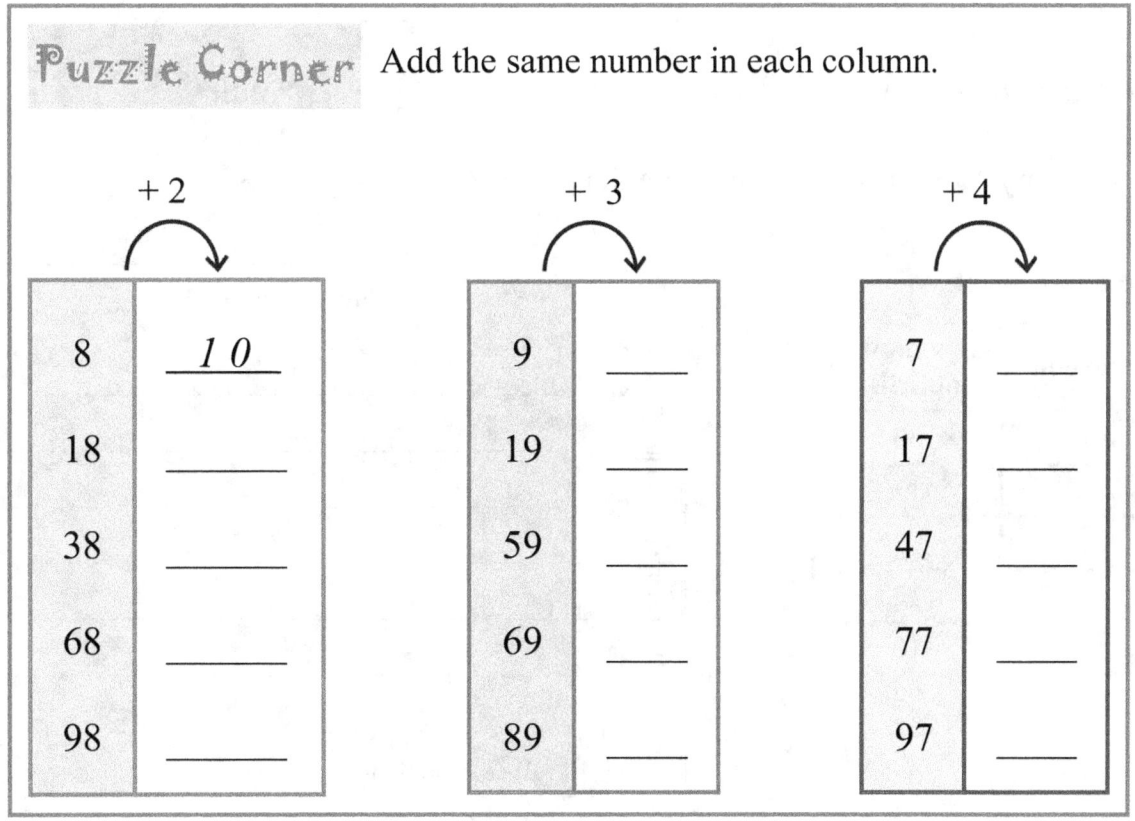

Puzzle Corner Add the same number in each column.

+2		+3		+4	
8	*1 0*	9	___	7	___
18	___	19	___	17	___
38	___	59	___	47	___
68	___	69	___	77	___
98	___	89	___	97	___

Pictographs

1. This is a **pictograph.** It shows how many miles each boy has ridden on his bike.
   Each bicycle picture means **10 miles**. A half-bicycle would be half that.

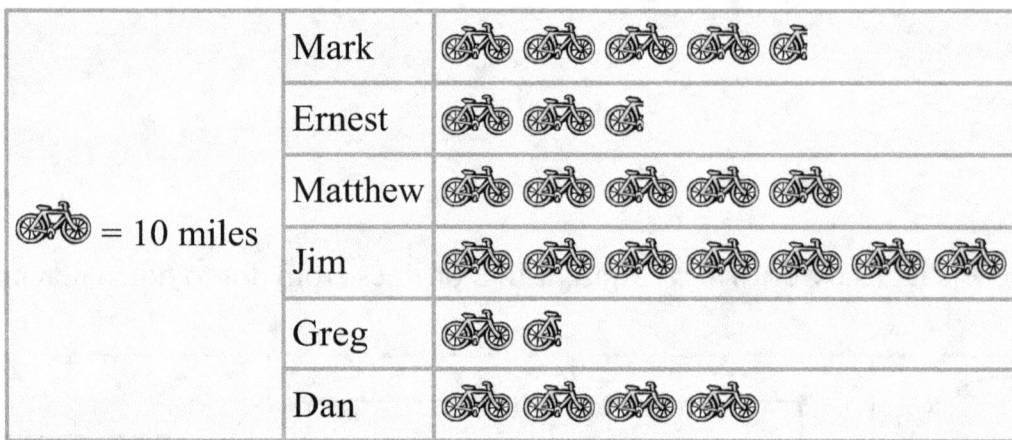

   a. Who rode the most miles? _____

    How many miles? _____ miles

   b. The boys who rode the fewest miles were Greg and _____.

    How many miles did they ride? _____ and _____ miles

   c. How many more miles did Matthew ride than Dan? _____ miles

   d. How many more miles did Dan ride than Greg? _____ miles

2. Make a bar graph.

Favorite color	How many people like it
Blue	30
Green	17
Red	13
Purple	12
Black	10

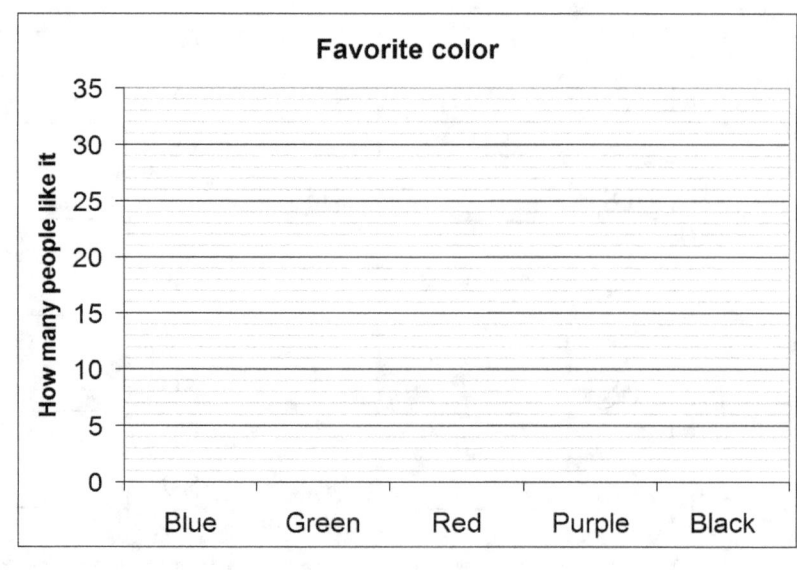

3. The children picked fruit from Grandpa Jerry's fruit trees. Each picture represents 6 pieces of fruit, so half a picture represents 3.

	How many?	
oranges		◯◯◖
mangos		〰〰〰◞
bananas		🍌🍌🍌🍌🍌

a. Fill in how many pieces of each kind of fruit they picked.

b. In total, how many oranges and bananas did they pick? You can use the grid on the right to add.

c. How many more bananas did they pick than mangos?

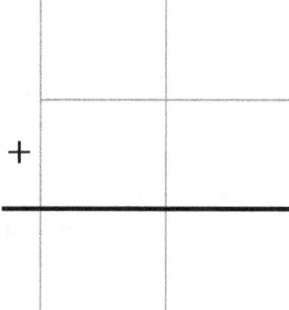

tens   ones

+

4. Some boys played a game of marbles and made a pictograph. Each marble is 5 points.

Mark	⦿⦿⦿⦿⦿
Aaron	⦿⦿⦿⦿⦿⦿⦿⦿
Henry	⦿⦿⦿⦿⦿
Jack	⦿⦿⦿⦿

Write three questions that you could ask another first grader about this pictograph. Then, ask the questions of your classmate or a friend. Check their answers.

Review Chapter 7

1. Solve. Write the letter after each problem in the purple box below the right answer.

60 + ____ = 68 A	90 − 60 = ____ D	52 + ____ = 58 L
52 + ____ = 55 U	60 − 8 = ____ T	24 + ____ = 26 H
22 + ____ = 27 O	91 − 20 = ____ M	55 + ____ = 56 O
43 + ____ = 50 O	32 − 10 = ____ N	11 + ____ = 20 U
56 − 6 = ____ E 80 − 20 = ____ L	82 − 10 = ____ F 100 − 4 = ____ Y	88 − 10 = ____ F

96   7   9   72   1   3   22   30   8   6   60   5   78   52   2   50   71

☐☐☐ ☐☐☐☐☐ ☐☐☐ ☐☐ ☐☐☐☐☐ !

2. Write the second number under the first and then add or subtract them.

a. 31 + 45   b. 70 + 19   c. 26 + 73   d. 31 + 8

e. 77 − 22   f. 56 − 14   g. 99 − 45   h. 47 − 5

130

3. **a.** Fill in the doubles chart at the right.

   b. Use the doubles chart to help you solve these
       addition problems. *Explain* how it helps you!

$7 + 8 =$ _____ $6 + 7 =$ _____

$6 + 5 =$ _____ $8 + 9 =$ _____

$5 + 5 =$ _____

$6 + 6 =$ _____

$7 + 7 =$ _____

$8 + 8 =$ _____

$9 + 9 =$ _____

4. Add. Tell which idea you use to add.

| Trick with nine | | Doubles chart |

a. $9 + 9 =$ _____   **b.** $8 + 4 =$ _____

| Trick with eight | | "Just one more" than a double |

c. $9 + 5 =$ _____   **d.** $7 + 7 =$ _____

| "Just one more" than a sum with 10 | | I just know it! |

e. $7 + 8 =$ _____   **f.** $6 + 5 =$ _____

g. $3 + 9 =$ _____   **h.** $6 + 7 =$ _____

5. Subtract.

a.
$11 - 2 =$ _____
$11 - 4 =$ _____
$11 - 5 =$ _____
$11 - 6 =$ _____

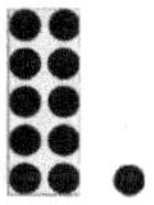

b.
$12 - 4 =$ _____
$12 - 5 =$ _____
$12 - 3 =$ _____
$12 - 6 =$ _____

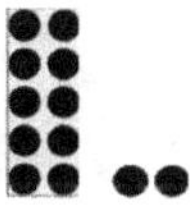

c.
$13 - 5 =$ _____
$13 - 6 =$ _____
$13 - 4 =$ _____
$13 - 7 =$ _____

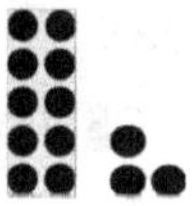

d.
$14 - 5 =$ _____
$14 - 8 =$ _____
$14 - 7 =$ _____
$14 - 6 =$ _____

e.
$15 - 6 =$ _____
$15 - 9 =$ _____
$15 - 7 =$ _____
$15 - 8 =$ _____

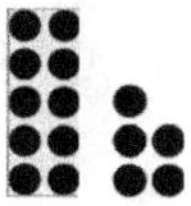

f.
$16 - 8 =$ _____
$16 - 9 =$ _____
$16 - 7 =$ _____
$16 - 6 =$ _____

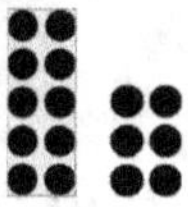

6. One book in the pictogram means that the child read <u>5 books</u>.

    a. How many books did
       Mariana read?

    b. How many books did
       Jose read?

    c. How many more books did
       Janet read than Jim?

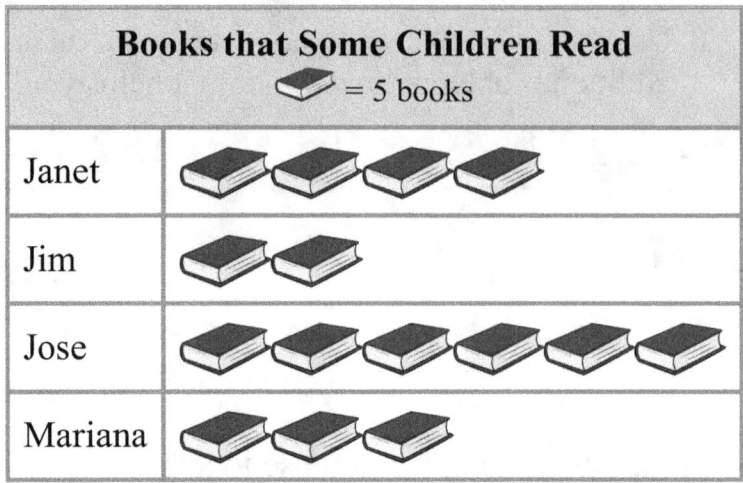

    d. How many more books did Jose read than Janet?

    e. Make your own question about the pictogram, and answer it!

7. Solve the word problems. Write number sentences for them.

a. Twenty birds were in a tree. Soon two flew away. Then five more flew away. How many were left?
b. Jack had five books. Then his mom gave him one more, and dad gave him three more. How many books does Jack have now?
c. Your sister is on page 14 of a 20-page book. How many pages does she have left to read?
d. How many years older is Sam than Jack, if Sam is 12 years old and Jack is 4?
e. Can you buy a \$15 train if you first have \$11 and then Mom gives you \$5?

8. Add. In some of these problems you need to make a new ten with some of the little dots. You can also use an abacus.

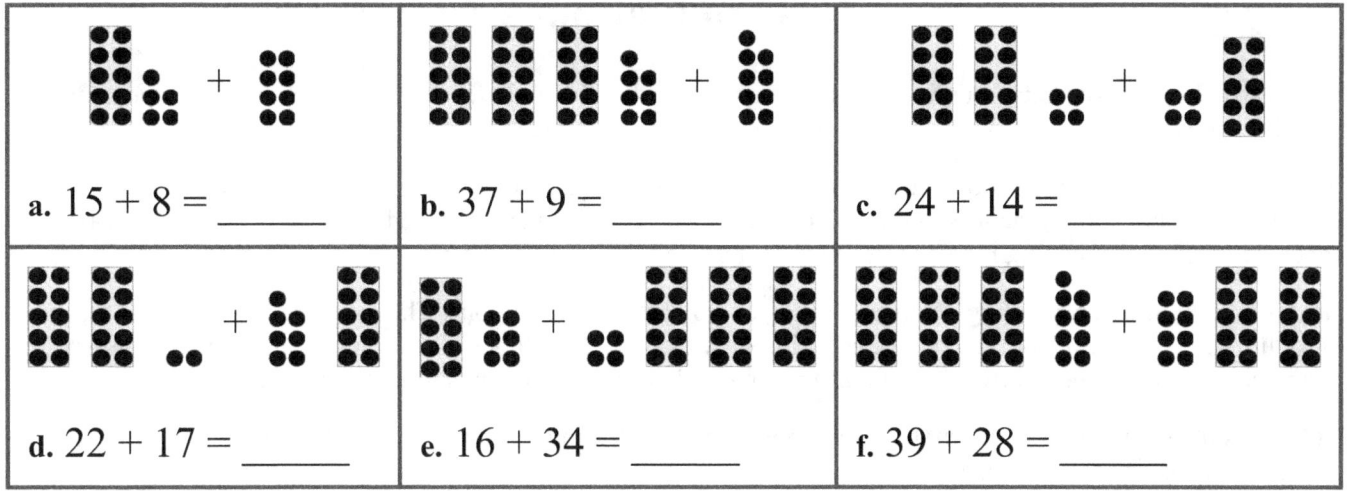

a. 15 + 8 = _____	b. 37 + 9 = _____	c. 24 + 14 = _____
d. 22 + 17 = _____	e. 16 + 34 = _____	f. 39 + 28 = _____

9. Pyramid numbers. Add two numbers that are next to each other, and put the sum *below* them both, in the middle.

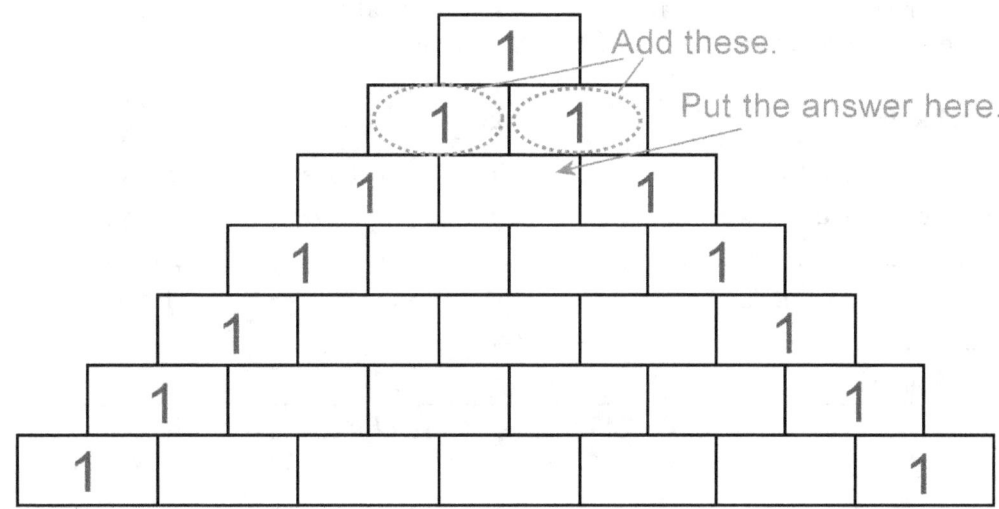

Add these.

Put the answer here.

Puzzle Corner Find what was subtracted!

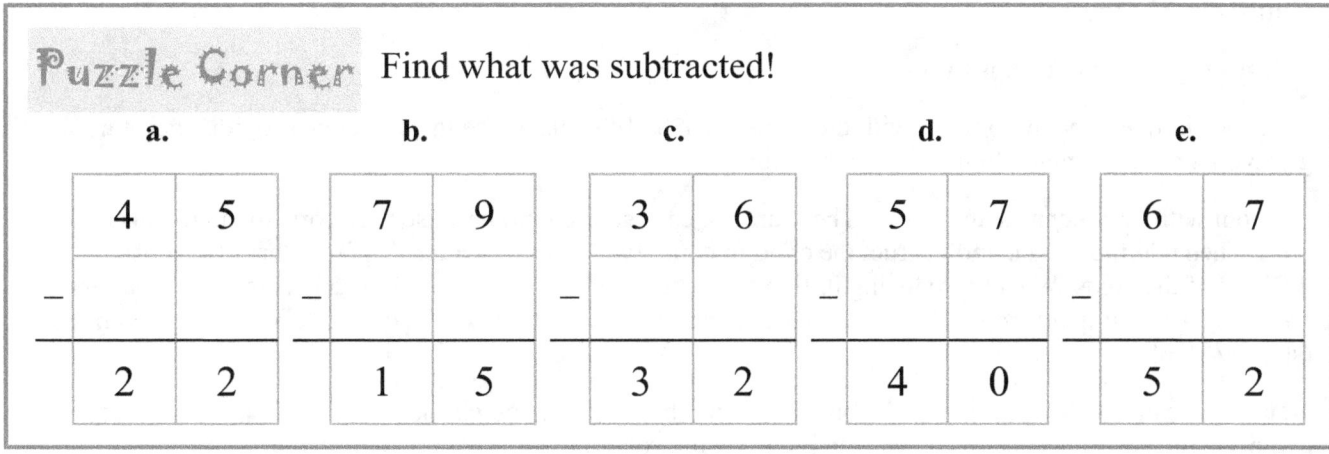

a.

4	5
−	
2	2

b.

7	9
−	
1	5

c.

3	6
−	
3	2

d.

5	7
−	
4	0

e.

6	7
−	
5	2

Chapter 8: Coins
Introduction

In this last chapter, we study counting coins. The goals are that the child learns to identify and count pennies, nickels, dimes, and quarters, when the amounts are 100 cents at most.

In the first lesson, we start out by counting only dimes and pennies, which is identical to practicing place value with tens and ones. The same lesson introduces the nickel. The child is instructed to count two nickels as 10 cents, which makes counting many coins much easier.

Next, children practice counting pennies, nickels, and dimes for two lessons. The following lesson then introduces the quarter. First, we practice counting only quarters and dimes, then quarters and nickels, and lastly all the coins. If counting quarters is difficult for the child, you can delay this topic till second grade.

Note: If you have the grayscale version of the book, or have printed in grayscale, it is helpful to color the pennies orange before doing the exercises.

Pacing Suggestion for Chapter 8

Please add one day to the pacing for the test if you will use it. Note that the specific lessons in the chapter can take several days to finish. They are not "daily lessons." As a general guideline, first graders should finish 1-2 pages daily or 7-9 pages a week. Please also see the user guide at https://www.mathmammoth.com/userguides/ .

The Lessons in Chapter 8	page	span	suggested pacing	your pacing
Counting Dimes, Nickels, and Cents	136	*3 pages*	2 days	
Counting Dimes, Nickels, and Cents 2	139	*2 pages*	2 days	
Quarters	141	*3 pages*	2 days	
Practicing with Money	144	*2 pages*	1 day	
Review - Coins	146	*1 page*	1 day	
Chapter 8 Test (optional)				
TOTALS		*11 pages*	8 days	

Games and Activities

Counting Money

You need: A bunch of coins to count.

Give the child an amount to make with the coins, such as 14 cents. Once the child does so, it is their turn to give you a money amount to make with the coins.

Start out with only pennies and dimes. These are the easiest to count, because they correspond to tens and ones. Then add the nickel, and instruct the child to count two nickels as a ten. The quarter is the most difficult of the coins. When introducing it, first teach the child that two quarters is 50 cents, three quarters is 75 cents, and four quarters is 100 cents. After that, you can go on to mixtures of quarters and other coins (step-by-step!).

Note: You can ask the child to check your work, and then in turn, you check theirs. In the course of the activity, you can then sometimes make an intentional error, so that the child can discover it.

Shopping Game

You need: Various items to purchase at the store, paper, pen, coins, a bag or wallet to keep money in.

Make a play store that has various items to purchase. The prices need to be less than one dollar. The child or student may enjoy choosing prices, and/or writing price tags for them. Typically, it is best if the teacher is the storekeeper, at least in the beginning.

In first grade, children may not understand about change, so at first, the idea is that the child will shop in the store, and pay with exact change. However, with time you might be able to introduce the idea of paying with a larger coin than what the item costs, and receiving change as the difference between what the item costs and what the customer pays.

Some children may enjoy it if the storekeeper writes a receipt for every purchase. After a while, you might switch roles and let the child be a storekeeper (and possibly write receipts). All of my children enjoyed this activity very much.

Games and Activities at Math Mammoth Practice Zone

Shopping Game
Practice making money amounts with coins and bills in this online game. You're shown an item to buy, and you click on coins/bills to make that exact amount.
https://www.mathmammoth.com/practice/shopping-game

Counting Money Game
Practice counting coins and bills in this online game! You can choose the exact coins and bills to use, the maximum for the total amount, the maximum number of coins/bills, and more. This allows you to make the activity as easy or difficult as desired.
https://www.mathmammoth.com/practice/count-money

Further Resources on the Internet

These resources match the topics in this chapter, and offer online practice, online games (occasionally, printable games), and interactive illustrations of math concepts. We heartily recommend you take a look. Many people love using these resources to supplement the bookwork, to illustrate a concept better, and for some fun. Enjoy!

https://l.mathmammoth.com/gr1ch8

Scan me

Counting Dimes, Nickels, and Cents

  This coin is called a cent or a penny. We write 1¢.

front   back

  This coin is called a dime. It is worth ten cents (10¢).

front   back

Count up to find how many cents there are in total:

10¢   20¢   21¢   22¢   23¢   24¢

Two dimes is 20¢.
Four pennies is 4¢.
The total is 24 cents.

1. Count and write the total amount in cents.

a.   _____ ¢

b.     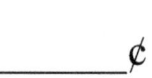 _____ ¢

c.        _____ ¢

d.  _____ ¢

e.  _____ ¢

f.  _____ ¢

g.   _____ ¢

2. Use real money to make these amounts. Or, draw gray circles with "10" for dimes, and orange circles with "1" for pennies.

a. 12¢	b. 40¢
c. 24¢	d. 31¢

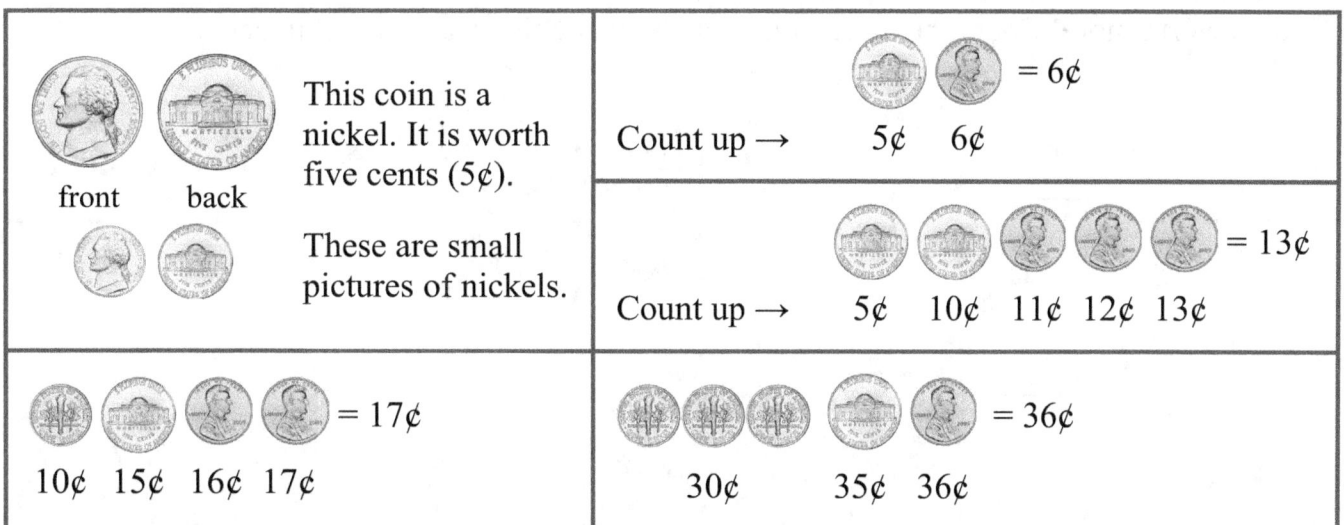

This coin is a nickel. It is worth five cents (5¢).

front   back

These are small pictures of nickels.

= 6¢

Count up →   5¢   6¢

= 13¢

Count up →   5¢   10¢   11¢   12¢   13¢

= 17¢

10¢   15¢   16¢   17¢

= 36¢

30¢   35¢   36¢

3. Find the coin value in cents.

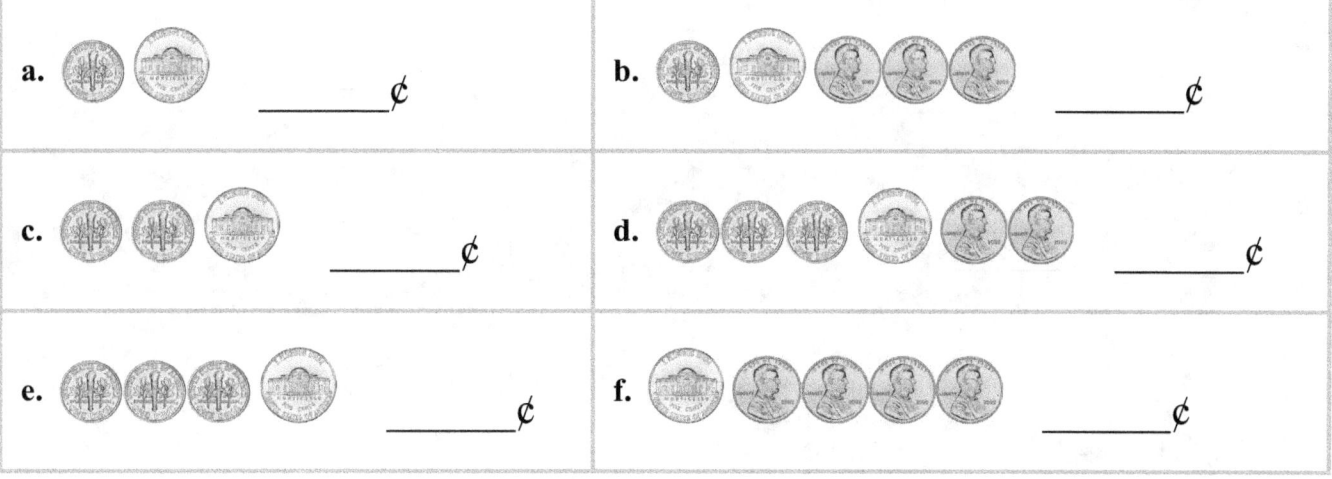

a. _____¢

b. _____¢

c. _____¢

d. _____¢

e. _____¢

f. _____¢

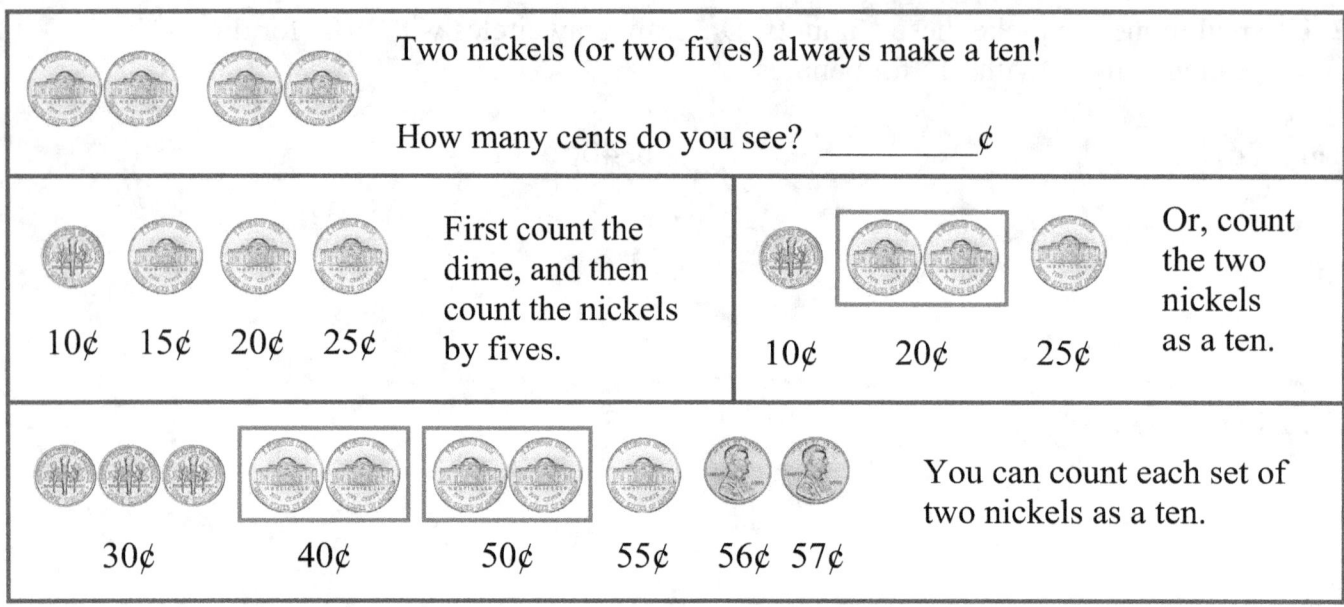

Two nickels (or two fives) always make a ten!

How many cents do you see? _____¢

First count the dime, and then count the nickels by fives.

10¢ 15¢ 20¢ 25¢

Or, count the two nickels as a ten.

10¢   20¢   25¢

You can count each set of two nickels as a ten.

30¢   40¢   50¢   55¢   56¢ 57¢

4. Dimes and nickels are sometimes hard to tell apart. A dime is a little *smaller* in size, but is worth more! Count the dimes and nickels. Write the total amount in cents.

a.  _____¢

b.  _____¢

c.  _____¢

d.  _____¢

e.  _____¢

f.  _____¢

g. 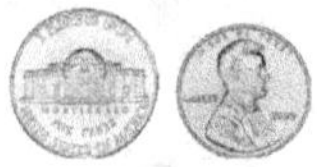 _____¢

h.  _____¢

i.  _____¢

j.  _____¢

k.  _____¢

l.  _____¢

Counting Dimes, Nickels, and Cents 2

1. Write the total amount in cents.

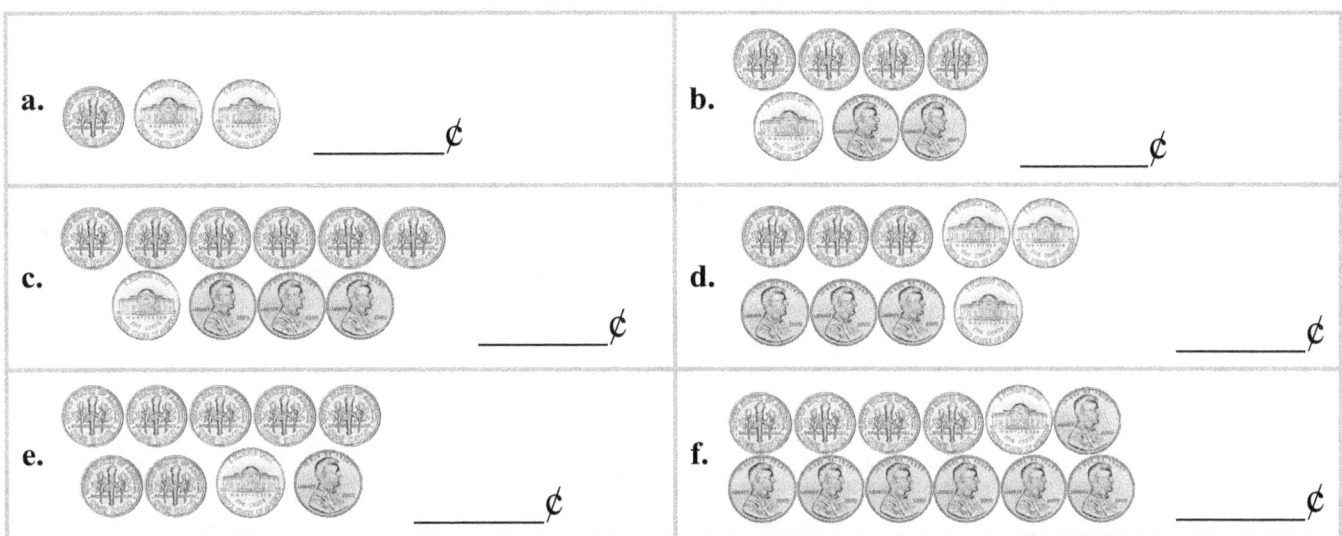

a. _____ ¢

b. _____ ¢

c. _____ ¢

d. _____ ¢

e. _____ ¢

f. _____ ¢

2. Draw one nickel more — how much money now?

a. _____ ¢

b. _____ ¢

c. _____ ¢

d. _____ ¢

e. _____ ¢

f. _____ ¢

3. Draw one dime more — how much money is there now?

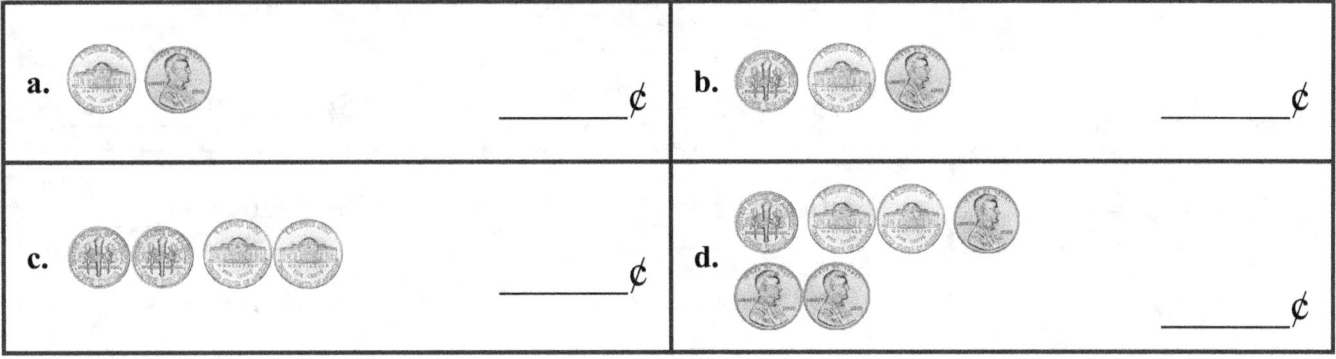

a. _____ ¢

b. _____ ¢

c. _____ ¢

d. _____ ¢

139

4. Use either real money, or draw gray circles with "10" to represent dimes, gray circles with "5" to represent nickels, and orange circles with "1" to represent pennies.

a. 25¢	**b.** 39¢	**c.** 14¢
d. 38¢	**e.** 63¢	**f.** 16¢
g. 61¢	**h.** 45¢	**i.** 27¢

5. You have some money, and then you get some more. Use real money or draw pictures to help.

a.

10¢ + 10¢ = _____ ¢

11¢ + 10¢ = _____ ¢

13¢ + 10¢ = _____ ¢

15¢ + 10¢ = _____ ¢

16¢ + 10¢ = _____ ¢

b.

21¢ + 5¢ = _____ ¢

24¢ + 5¢ = _____ ¢

25¢ + 5¢ = _____ ¢

20¢ + 5¢ = _____ ¢

27¢ + 5¢ = _____ ¢

c.

40¢ + 20¢ = _____ ¢

53¢ + 10¢ = _____ ¢

55¢ + 5¢ = _____ ¢

56¢ + 20¢ = _____ ¢

58¢ + 30¢ = _____ ¢

Quarters

One **quarter** is 25 cents.

The word "quarter" means one-fourth. A quarter coin is one-fourth part of a **dollar**.

One dollar is 100 cents, and is written $1.

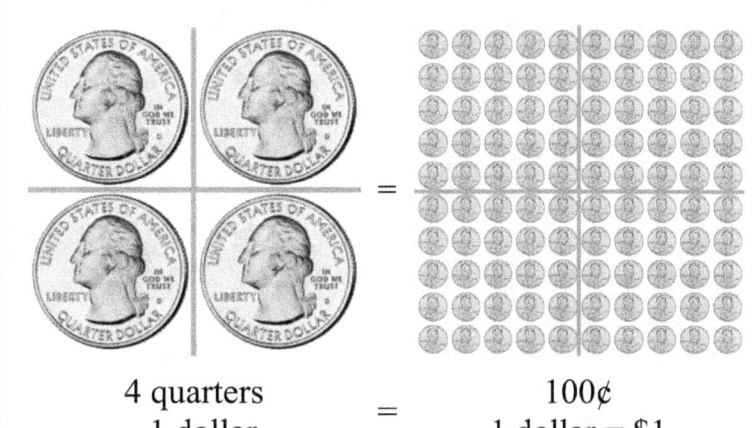

4 quarters		100¢
1 dollar	=	1 dollar = $1

 Two quarters = 50¢.

 Three quarters = 75¢

25¢   35, 45, 55   56, 57¢   Count the quarters first since
       (count dimes by tens)   they have the biggest cent-value.

1. Quarters and dimes. Write the total amount in cents.

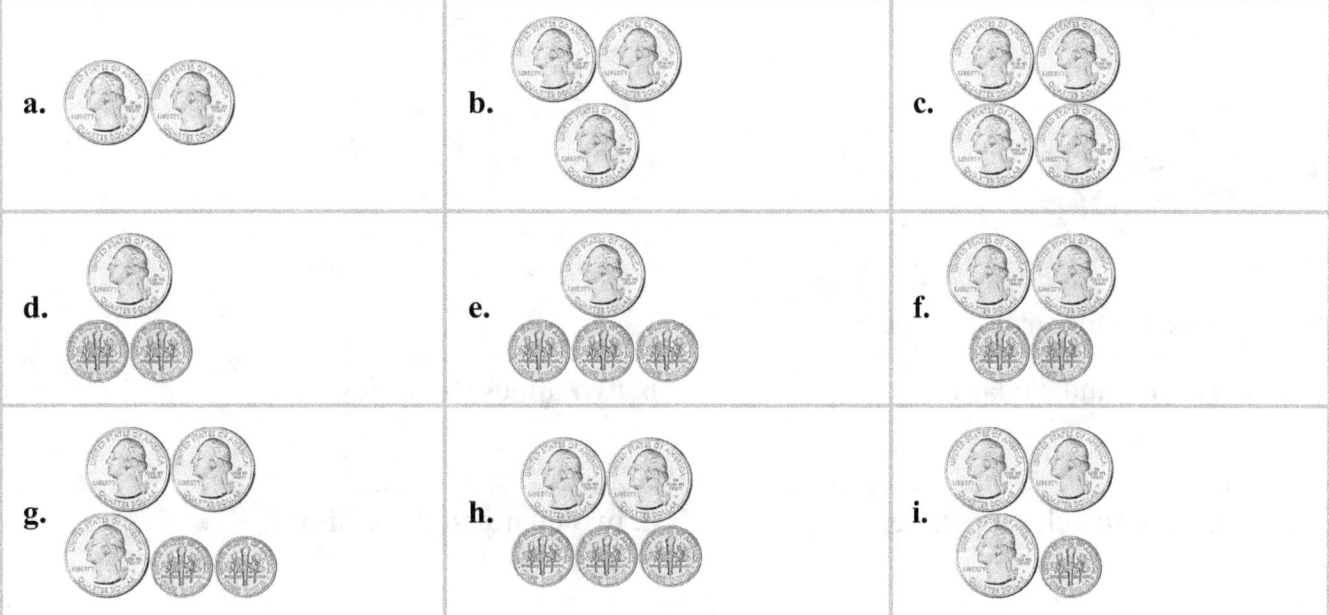

2. Quarters and nickels. Write the total amount in cents.

3. How much money? Write down the amount in cents.

a. 	b. 
c. 	d. 
e. 	f. 
g. 	h. 
i. 	j. 
k. 	l. 

4. How much is the total if you have:

a. two dimes and a quarter	**b.** two dimes, four nickels
c. a dime, a nickel, six pennies	**d.** two quarters, three dimes, seven pennies

142

5. Cross out the coins you need to buy the item. Write how many cents you have left.

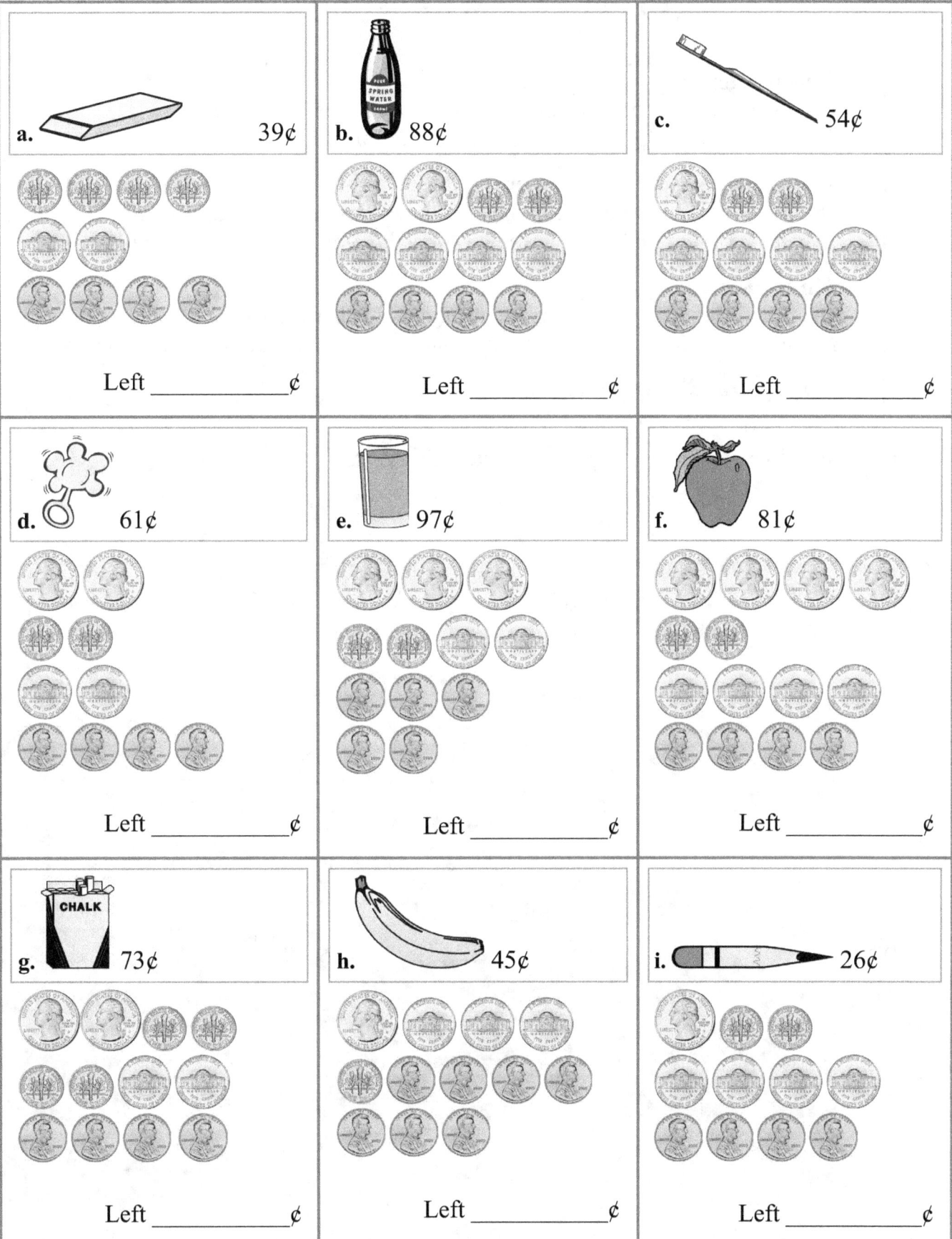

Practicing with Money

- **One quarter** = _____ cents.
  Use ONE quarter when the money amount is between 25 and 50 cents.
  Example: To make 31 cents, take one quarter, one nickel, and one penny.

- **Two quarters** = _____ cents.
  Use TWO quarters when the money amount is between 50 and 75 cents.
  Example: To make 62 cents, take two quarters, one dime, and two pennies.

- **Three quarters** = _____ cents.
  Use THREE quarters when the money amount is between 75 and 100 cents.
  Example: To make 87 cents, use three quarters, one dime, and two pennies.

- **Four quarters** = 100 cents or one dollar.

1. Draw the coins you would use to pay for an item that costs:

You have:

a. 29¢	b. 46¢	c. 62¢
d. 48¢	e. 86¢	f. 91¢

In the following exercises, either use real money, or draw to illustrate:
- orange circles with "1" for pennies.
- gray circles with "5" for nickels
- gray circles with "10" for dimes
- a little bigger gray circles with "25" for quarters

2. Illustrate these amounts of money. Use one quarter in each problem.

a. 30¢	b. 32¢	c. 35¢
d. 45¢	e. 41¢	f. 48¢

3. Illustrate these amounts of money. Use two quarters in each problem.

a. 50¢	b. 53¢	c. 58¢
d. 60¢	e. 66¢	f. 72¢

Review - Coins

1. How much money? Write down the amount in cents.

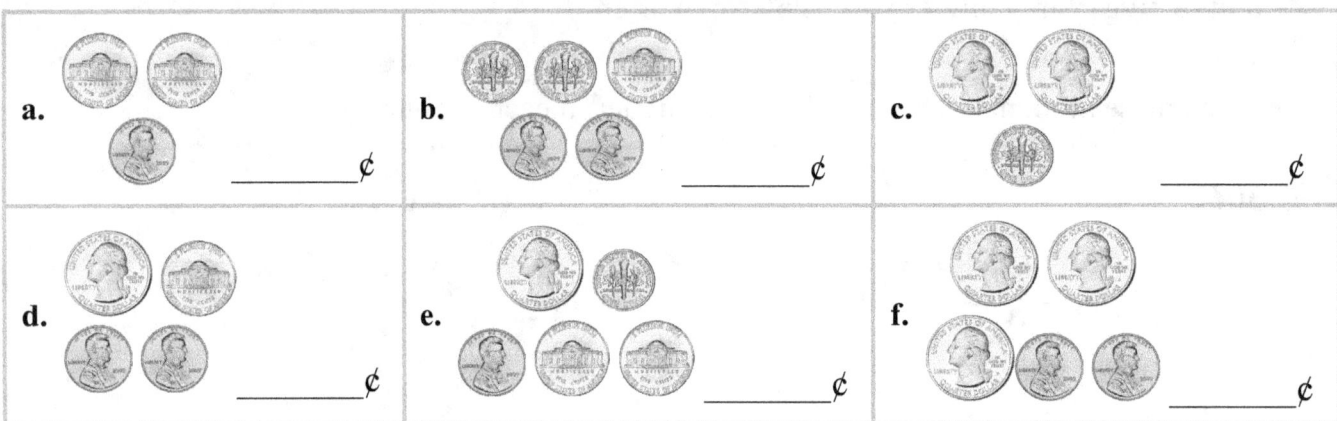

a. _____ ¢	b. _____ ¢	c. _____ ¢
d. _____ ¢	e. _____ ¢	f. _____ ¢

2. Draw coins to illustrate these amounts of money.

a. 52¢	b. 27¢	c. 76¢
d. 85¢	e. 79¢	f. 34¢

3. You buy an item. How much money will you have left?

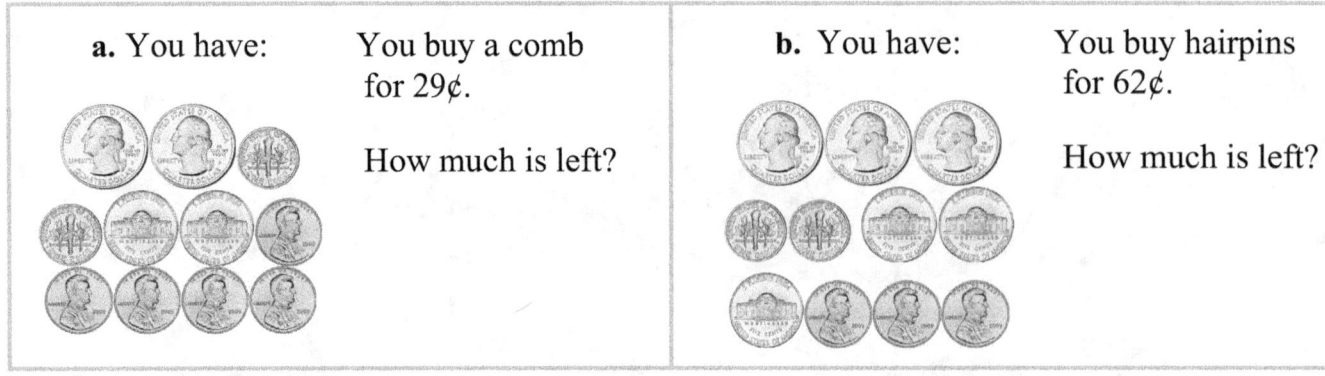

a. You have:	You buy a comb for 29¢. How much is left?	**b.** You have:	You buy hairpins for 62¢. How much is left?